T0144488

Smart Embedded Systems and Applications

RIVER PUBLISHERS SERIES IN ELECTRONIC MATERIALS, CIRCUITS AND DEVICES

Series Editors

JAN VAN DER SPIEGEL
University of Pennsylvania,
USA

MASSIMO ALIOTO
National University of Singapore,
Singapore

KOFI MAKINWA
Delft University of Technology,
The Netherlands

DENNIS SYLVESTER
University of Michigan,
USA

MIKAEL ÖSTLING
KTH Stockholm,
Sweden

ALBERT WANG
University of California,
Riverside, USA

Indexing: all books published in this series are submitted to the Web of Science Book Citation Index (BkCI), to SCOPUS, to CrossRef and to Google Scholar for evaluation and indexing. All River Publishers books in the area are available on the IEEE Explore platform.

The "River Publishers Series in Electronic Materials, Circuits and Devices" is a series of comprehensive academic and professional books which focus on theory and applications of advanced electronic materials, circuits and devices. This includes analog and digital integrated circuits, memory technologies, system-on-chip and processor design. Also theory and modeling of devices, performance and reliability of electron and ion integrated circuit devices and interconnects, insulators, metals, organic materials, micro-plasmas, semiconductors, quantum-effect structures, vacuum devices, and emerging materials. The series also includes books on electronic design automation and design methodology, as well as computer aided design tools.

Books published in the series include research monographs, edited volumes, handbooks and textbooks. The books provide professionals, researchers, educators, and advanced students in the field with an invaluable insight into the latest research and developments.

Topics covered in this series include:-

- Analog Integrated Circuits;
- Data Converters;
- Digital Integrated Circuits;
- Electronic Design Automation;
- Insulators;
- Integrated circuit devices;
- Interconnects;
- Memory Design;
- MEMS;
- Nanoelectronics;
- Organic materials;
- Power ICs;
- Processor Architectures;
- Quantum-effect structures;
- Semiconductors;
- Sensors and actuators;
- System-on-Chip;
- Vacuum devices.

For a list of other books in this series, visit www.riverpublishers.com

Smart Embedded Systems and Applications

Editor

Saad Motahhir

École Nationale des Sciences Appliquées,
Sidi Mohamed Ben Abdellah University, Morocco

River Publishers

Routledge
Taylor & Francis Group
LONDON AND NEW YORK

Published 2022 by River Publishers

River Publishers

Alsbjergvej 10, 9260 Gistrup, Denmark

www.riverpublishers.com

Distributed exclusively by Routledge

4 Park Square, Milton Park, Abingdon, Oxon OX14 4RN

605 Third Avenue, New York, NY 10017, USA

Smart Embedded Systems and Applications / Saad Motahhir.

Routledge is an imprint of the Taylor & Francis Group, an informa business

ISBN 978-87-7022-772-8 (print)

ISBN 978-10-0084-966-0 (online)

ISBN 978-1-003-37569-2 (ebook master)

While every effort is made to provide dependable information, the publisher, authors, and editors cannot be held responsible for any errors or omissions.

I am honored to dedicate this book to my parents and my
Prof. Abdelaziz El Ghzizal.

Contents

Preface

An embedded system is a combination of computer hardware and software designed for a specific function. Embedded systems smartly integrate hardware and software components. They sometimes are organized as systems of systems. They usually are massively connected. They often are safety-critical. Applications of embedded systems are generally found in image processing, Industry 4.0, medical, aerospace, automotive, military applications, etc., such as autonomous vehicles, computer vision, modern drones, within telecommunications networks. Hence, embedded systems designed for the above-mentioned applications must be smart. Moreover, the quality of embedded systems is assessed by following specialized standards according to the application domain. For instance, Autosar, Misra C, Aspice, ISO26262 are followed in the automotive industry.

The design of such smart embedded systems requires advanced digital technologies, such as artificial intelligence, Internet of Things or blockchains, to integrate a variety of modern techniques in order to improve system performance in various applications.

This book covers a wide range of challenges, technologies and state-of-the-art for the design, development and realization of smart and complex embedded systems and their applications; i.e., software and hardware development, with the use of digital technologies, and quality assurance for critical applications.

This book starts with automotive safety systems which is one of the major functional domains. It discusses the importance of software in automotive systems followed by an insight into Automotive Software Standards, ISO26262, and Autosar. The book further discusses the use of processor in the loop test for an adaptive trajectory tracking control for quadrotor UAVs. It also illustrates the role of embedded systems in medical engineering. Various innovative applications involving the concept of image processing and Internet of Things are also presented in this book. The SoC power estimation is also investigated. Finally, a review of the hardware/software partitioning algorithms with some future works have been presented. This book is intended for academicians, researchers, and industrialists.

Saad Motahhir

Acknowledgments

This book could not be that successful without the effort of the authors and reviewers. Therefore, I would like to express my sincere appreciation to all of you who generously supported this book.

S. Motahhir

List of Reviewers

Abdellah El Kamili, *Lear technopolis, Rabat, Morocco*

Ahmed Gaga, *Faculté Polydisciplinaire, USMS, Benni Mellal, Morocco*

Wissam Jenkal, *ENSA, UIZ, Agadir, Morocco*

Mhamed Sayyouri, *ENSA, USMBA, Fez, Morocco*

Mustapha Errouha, *ENSEEIHT, University of Toulouse, France*

Nabil El Akkad, *ENSA, USMBA, Fez, Morocco*

Hajar Saikouk, *INSA, Euromed University of Fes (UEMF), Fez, Morocco*

Aboubakr El Hammoumi, *EST, USMBA, Fez, Morocco*

Moad Essabbar, *INSA, Euromed University of Fes (UEMF), Fez, Morocco*

Moussa Labbadi, *INSA Hauts-de-France, Université polytechnique Hauts-de-France, Valenciennes, France*

Abdelilah Chalh, *EST, USMBA, Fez, Morocco*

Ahmed Hanafi, *EST, USMBA, Fez, Morocco*

Claude Baron, *INSA Toulouse, France*

Anass Mansouri, *ENSA, USMBA, Fez, Morocco*

List of Figures

List of Tables

List of Notations and Abbreviations

Abbreviations	Definition
A2S	Automotive systems simulation
ACUA	Automotive Company Under Audit
AI	Artificial intelligence
ANNs	Artificial neural networks
APP	Acceleator pedal position
ASIC	Application specific integrated circuit
ASIL	Automotive safety integrity level
ATI	Accurate Technologies Inc.
AVL	Anstalt für Verbrennungskraftmaschinen List (Company)
BDC	Bottom dead centre
BMBF	Federal ministry of education and research
CA	Crank angle
CAN	Controller area network
CNNs	Convolutional neural networks
DAC	Digital-analogue converter
DL	Deep learning
DMA	Direct memory access
DSP	Digital signal processing
DT	Digital twin
DUT	Device under-test
ECU	Electronic control unit
FAO	Food and agriculture organization
FMU	Functional mock-up
FPGA	Field programmable gate array
FSA	Functional safety assessment
FS	Functional safety
HDL	Hardware description language
HIL	Hardware-in-the-loop
HLL	High-level language

HLS	High-level synthesis
HW-SW	Hardware-software
IEC	International electrotechnical commission
IOT	Internet of things
IPCC	The intergovernmental panel on climate change
IP	Intellectual property
ISO	International organization for standardization
LMS	Least mean square
M2M	Machine to machine
MAF	Mass air flow
MAP	Manifold air pressure
ML	Machine learning
MODIS	Moderate resolution imaging spectro-radiometer
MPL	Multi-layer perceptron
MPV	Multi-purpose vehicle
NDVI	Normalized difference vegetation index
NEDC	New european driving cycle
NICT	New information and communication technologies
NLMS	Normalized least mean square
OBD	On-board diagnostics
OEM	Original equipment manufacturer
PCB	Printed circuit board
PC	Personal computer
PMW	Pulse width modulation
RAM	Random access memory
RDE	Real driving emissions
ROTS	Real-time operating system
RPAS	Remotely piloted aircraft system
RTL	Register transfer level
SDK	Software development kit
SIL	Software-in-the-loop
SSCM	Site-specific crop management
SSSA	Secured smart sustainable agriculture
TDC	Top dead centre
UAV	Unmanned aerial vehicle
WLTC	Worldwide harmonized light vehicles test cycle

SECTION 1

Smart Embedded Systems for the Automotive Industry

1

Functional Safety Audit/Assessment for Automotive Engineering

A. El Kamili[1], Y. El Kharaz[2], A. Tribak[1]

[1]National Institute of Posts and Telecommunications, Rabat, Morocco
[2]University of sciences and technologies, FEZ, Morocco
[1]Elkamili@inpt.ac.ma, Tribak@inpt.ac.ma; [2]youssef.elkharaz@usmba.ac.ma

Abstract

Automotive safety is important because lives and reputations are at stake. In this chapter, we present an automotive overview of functional safety audits and assessments. The safety management activity of an organization needs a periodic check ensured by functional safety audit and assessment. In accordance with ISO26262, the automotive company should have a defined process for functional safety audit and assessment. In this overview, we present some basic definitions related to ISO26262 and ASIL, categories of audits and assessments, procedures, and phases to comply with safety requirements. In addition, we define the scope, objectives, roles and responsibilities, and departments inside the automotive company looking to ensure a well-defined process for functional safety audit and assessment.

1.1 Introduction and Objectives

Safety auditing and assessment is a mais safety management activitiy. Safety audits and assessments provide a means for systematically assessing how well the organization is achieving its safety goals. The Functional Safety Audit and Functional Safety Assessment (FSA) Program, together with other safety oversight activities, e.g. safety performance monitoring, provides feedback to managers concerning the safety performance of an organization and/or a specific development/project.

An audit can be defined as "A systematic, independent and documented process for obtaining audit evidence and evaluating it objectively to determine the extent to which audit criteria are fulfilled." Similarly, a functional safety audit can be defined as a "systematic and independent examination to determine whether the procedures specific to the functional safety requirements comply with the planned arrangements, are implemented effectively, and are suitable to achieve the specified objectives." or defined as "A proactive safety management activity providing a means of identifying potential problems before they have an impact on safety."

The functional safety audit procedure [2] described in the supporting paragraphs establishes the practices and methods by which automotive organization or companies will implement an effective functional safety audit program and ensure that the implemented functional safety processes and procedures conform to all applicable safety standards, specifically ISO 26262 [8] [9]

The FSA Procedure described in the supporting sub-pages establishes the practices and methods by which the automotive company will carry out effective FSA against specific projects, in compliance with ISO 26262.

The functional safety audit's purpose is to determine the conformity or nonconformity with specified requirements related to functional safety processes and procedures. then determine the capability of a supplier and provide the auditee with an opportunity to improve the functional safety management system. In addition, it aims to assess adherence to regulatory requirements, e.g. to determine if the automotive organization implementation of functional safety management and associated procedures [4] is meeting the intent of ISO 26262 and assess adherence to requirements for certification to a management system or regulatory standard.

The functional safety assessment will determine the degree of conformance to and effectiveness of the overall functional safety development life cycle for specific projects and provide advice and recommendations for the improvement of the functional safety development life cycle for specific projects.

The procedure is applicable to all automotive organizations or companies having a functional safety team or department and the functional safety audit/assessment shall be applied at ASILs (B), C, or D.

The products covered by our functional safety audit and assessment include:

- Safeguards and safety components in machinery (e.g. electro-sensitive protective equipment).

- Programmable or configurable controllers with safety functions.

- Drive systems with safety functions.

- Bus systems, devices with safety-related bus communication.

- Furnaces, controls and safeguards for fuel / air.

- Safety related modules and components (e.g. relays with forcibly guided contacts, position switches, valves).

- Sensors e.g. for position, temperature, mass flow, filling level, pressure, and detection of gases.

- ASICs and FPGAs in safety-related applications.

- Software products (compiler, programming and configuration tools, operating systems, hypervisors, software stacks, or adaptation layers).

The following paragraphs will explore more details about functional safety audits and assessments in automotive embedded systems engineering.

1.2 ISO 26262 and ASIL Overview

ISO 26262 is a safety related standard that's derived from IEC 61508 [7]. It applies to electronic or/and electric systems in production vehicles. This includes driver assistance, propulsion, and vehicle dynamics control systems. The goal of the standard is to ensure safety throughout the lifecycle of automotive equipment and systems. ISO26262 contains 10 parts:

- Part 1: Vocabulary.

- Part 2: Management of functional safety.

- Part 3: Concept phase.

- Part 4: Product development at the system level.

- Part 5: Product development at the hardware level.

- Part 6: Product development at the software level.

- Part 7: Production and operation.

- Part 8: Supporting processes.

- Part 9: ASIL-oriented and safety-oriented analysis.

- Part 10: Guideline on the safety standard.

Automotive Safety Integrity Level (ASIL) is the main component of ISO 26262 and it is established by performing a risk analysis of a potential hazard by looking at the controllability, exposure, and severity of a specific component [6]. There are four Automotive Safety Integrity Levels, named

A–D. ASIL A dictates the lowest level of risk and ASIL D is the highest integrity requirements on the product, as you go from A to D, the compliance requirements get stricter.

1.3 Functional Safety Audit/Assessment Program

An FS Inspection provides a methodical and independent examination of the particular safety lifecycle phase conditions that misbehave with the planned arrangements, are enforced effectively, and are suitable for conditioning under review. It determines whether the "procedures" specific to functional safety admit the specified objects.

An FS Inspection is accepted to ensure compliance with procedures. Adjudicators don't assess the acceptability of the work they're auditing and don't make specific judgments about functional safety and integrity.

In addition, an FSA is an independent in-depth disquisition into the former and current lifecycle phase conditioning grounded on substantiation, aimed at assessing whether functional safety has been achieved. FSAs calculate heavily on assessor judgments and faculty. One of the inputs to the FSA process is the FS Inspection processes and findings.

There are three different categories of Functional Safety Audit/ Assessment program:

- The Internal Functional Safety Audit/Assessment: this may include a functional safety audit of documented functional safety processes and procedures independent of the project or/and a functional safety assessment against a specific project.

- External Functional Safety Audit/Assessment: this can include a Functional Safety Audit of ACUA Document Processes and Procedure or an FSA against a specific Project conducted by a third party (e.g. OEM FSA).

- Supplier Audit: this represents a Functional Safety Audit of an external supplier conducted by an automotive organization under audit, e.g., to assure they meet certain requirements for Functional Safety.

The trigger for a Functional Safety Audit and/or FSA is the schedule within the Functional Safety Audit/Assessment Program Schedule. This program schedule details the planned internal, external, and supplier Functional Safety Audits/Assessments. For the conduct of a Functional Safety Audit or FSA there are four basic phases, namely:

1.   Functional Safety Audit / FSA Planning.

2.   Preparing for the Functional Safety Audit / FSA Activities.

3.   Performing the Functional Safety Audit FSA.

4.   Reporting and Follow-up.

An example of an 'Automotive organization under Audit' Functional Safety Audit / FSA Program Procedure is shown in Figure 1.1.

1.3.1 Internal functional safety audit procedure

1.3.1.1 Definition and safety compliance procedures phases

The internal functional safety audit assesses the adequacy of the documented functional safety processes and conformance of these processes to the ISO 26262 standard as indicated in Figure 1.2. The functional safety audit, via independent review and examination of records and activities, assesses the adequacy of system controls, processes and procedures, to ensure compliance with established policies, standards and operational procedures, and recommends necessary changes in controls, policies, or procedures. Table 1.1 presents a tracking example of an internal safety audit inside an automotive company.

Internal Functional Safety Audit specifies the method for supervising and controlling functional safety procedures throughout the automotive company branch. It comprises two main activities: formal audits and regular and ad-hoc inspections.

The formal audits provide a more comprehensive and formal audit towards the compliance of functional safety procedures and plans to ISO 26262. They are established as per the following process lifecycle illustrated in Figure 1.3:

- Starting audit defined as Phase1 enrolling only the main topics from ISO26262.

- Completion audit defined as Phase 2 comprising the outputs review from Phase1 and the rest of ISO26262 requirements. These two steps form an audit and need to be scheduled within a maximum of 1 year in between, depending on the gap identified.

- Formal regular audit defined as maintenance including the complete set of ISO26262 in a single shot.

    Phase 1 will include:

- ISO26262 works products from Parts 2, 4, 5, 6, and 8.

- ISO26262 Part2 requirements.

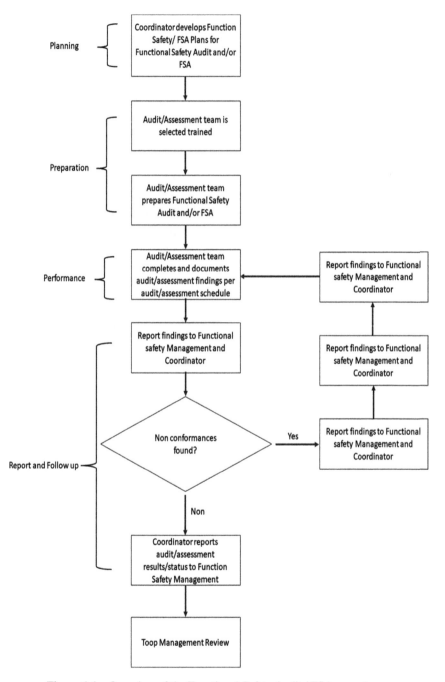

Figure 1.1   Overview of the Functional Safety Audit / FSA procedure.

Table 1.1 Example of an internal safety audit inside an automotive company.

Procedure title	Procedure reference	Version	Date	Next review date	Procedure owner
Internal FS Audit	AUT-C-FS-P-IFA	V0.1	October 2021	April 2022	Functional safety team

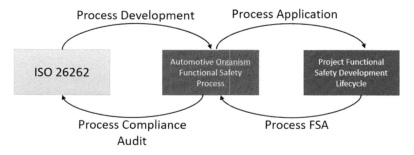

Figure 1.2 Process conformance and compliance with ISO26262.

- ISO26262 Part8 requirements from chapters:
  - ○ Specification and management of safety requirements
  - ○ Configuration management
  - ○ Change management
  - ○ Verification
  - ○ Documentation

  Phase 2 will include:
- ISO26262 works products from Parts 2, 4, 5, 6, and 8.
- ISO26262 Part2 requirements.
- ISO26262 Part8 requirements from chapters:
  - ○ Interfaces within distributed developments
  - ○ Specification and management of safety requirements
  - ○ Configuration management
  - ○ Change management
  - ○ Verification

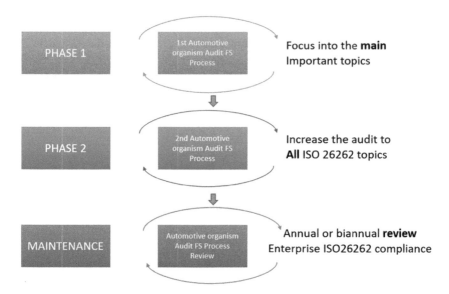

Figure 1.3 Phases of compliance with functional safety procedures and plans to ISO 26262.

  ○ Documentation

  ○ Confidence in the use of software tools

  ○ Qualification of software components

  ○ Qualification of hardware components

  ○ Proven in using argument

• ISO26262 Method Selection Guidelines for parts 4, 5, 6, and 8 as per the criteria: highly recommended (++) for ASILC

• ISO26262 Part4, Part5, Part6 requirements.

  The maintenance phase will include:

• ISO26262 WP from Parts 2, 4, 5, 6 and 8.

• ISO26262 Part2 requirements.

• ISO26262 Part8 requirements from chapters:

  ○ Interfaces within distributed developments

  ○ Specification and management of safety requirements

○ Configuration management

○ Change management

○ Verification

○ Documentation

○ Confidence in use of software tools

○ Qualification of software components

○ Qualification of hardware components

○ Proven in use argument

- ISO26262 Methods and Guidelines(specific to each automotive entity) for parts 4, 5, 6 and 8 as per the criteria: highly recommended (++) for ASILC

- ISO26262 Part7, Part9, Part10.

1.3.1.2 Objective and scope of internal FS audit

The objective of the automotive company Internal FS audit is to provide management with an independent assessment of the capability progress, quality, and attainment of project objectives based on company policies and procedures. Then to provide management with an assessment of the adequacy of Internal Functional Safety Management processes and that the processes are applied consistently across all projects. Another objective is to provide management with an evaluation of the internal controls of proposed Functional Safety processes at a point in the development cycle where enhancements can be easily implemented and processes adapted, in addition to checking if an assessment of the adequacy of Functional Safety controls is implemented and provide an evaluation of the enterprise Functional Safety metrics.

The main purpose of an internal FS audit is to determine if the enterprisc's Functional Safety is on-track or not. If it is off track, we determine why and offer specific advice on how to get it back on track. Finally, to determine whether what causes the issues are systemic and how to avoid them in the future.

The scope concerns the functional safety requirements from ISO26262 including methods and guidelines, work products, and requirements from the following ISO26262 chapters: Ch1 to Ch10.

This procedure is applicable to the complete automotive company or branch side and applied at ASILs (B), C, and D.

In relation to the internal FS audit we find the ISO 26262 requirements described in ISO 26262 Standard, Part 2 (section 6.4.8 and 6.4.9)

1.3.1.3 Internal functional safety audit procedure:

To put in place an efficient procedure for an internal functional safety audit, we define the roles and responsibilities of stakeholders as well as the concerned departments and entities by the audit. The definition of the procedure is a key element in the audit, everyone should know what to do and how to do it according to clear planning and selection.

> **Overall Responsibility:** Safety Manager.
>
> **IFA Responsible:** Functional Safety Auditor Responsible (selected from the automotive company or entity functional safety Team as appropriate).
>
> **IFA Initiation (when):** in the second quarter, Q2, 2021 as the initiation phase. Maximum Q1 2022 as completion phase. Once every two years on a regular basis and according to the Functional Safety Department timing.
>
> **IFA Input(s)/Resources:** access to enterprise staff, documentation, and work products.

IFA Roles and Responsibilities:

- **Safety Manager**: Report to management the findings of the audit.

- **Functional Safety Team:** Plan, prepare and conduct the IFA. Prepare and issue the IFA report. Follow-up on any non-conformance reports issued throughout the audit. Report to the Safety Manager the findings of the audit. Escalate any non-conformance as necessary to the Safety Manager.

- **Functional Safety Auditor Responsible:** Provide and arrange access to internal personnel and documentation. Action any Nonconformance reports issued during the audit.

Project FSA Procedural Steps:

- Planning

- Preparation for audit

- Auditor selection

- Conduction of audit

- Audit report and recommended actions

- Corrective actions and priorities

Audited Departments:

> **Management**
* Responsible: Safety manager

> **Systems**
* Responsible : Systems manager or System process responsible

> **Hardware**
* Responsible : Hardware manager or hardware process responsible

> **Software**
* Responsible : SW manager or SW process responsible

> **Supporting Processes**
* Responsible : Processes manager

Functional Safety Auditor selection:

The auditor/(s) selection is done through an internal accreditation based on the candidate's background, training, and experience.

A specific process of accreditation with documentation should have a place in the automotive entity or department.

1.3.1.4 Non-Conformance

A non-conformance as defined by ISO is "The non fulfillment of specific requirements." "The definition covers the departure or absence of one or more quality characteristics or quality system elements from specified requirements. A non-conformance may be a failure to:

* Comply with the applicable standard.

* Implement quality manual, procedures, or other documentation requirements specified by the automotive company.

* Implement a code of practice, regulation, contract..

1.3.2 Internal functional safety assessment procedure

The internal safety assessment [1] evaluates the adequacy of the project's Functional Safety Development and conformance to the documented Functional Safety Process.

FSA project specifies the method for assessing, supervising, and controlling Functional Safety procedures and activities through the project lifecycle. It comprises one main activity performed several times throughout the project lifecycle.

Table 1.2   Example of an internal safety assessment inside an automotive company.

Procedure title	Procedure reference	Version	Date	Next review date	Procedure owner
Internal FS Assessment	AUT-C-FS-A-IFA	V0.1	Octobre 2021	April 2022	Functional safety team

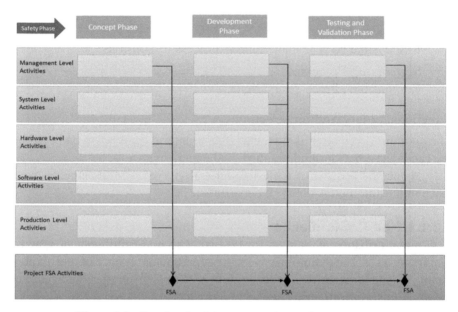

Figure 1.4   Functional safety assessment procedure.

The formal assessments provide a more comprehensive and formal assessment of compliance with Functional Safety Procedures. They should be carried out at key points during the project life cycle, e.g. at the concept phase, development phase, and validation phase.

1.3.2.1 Objective and scope of internal FS assessment

The objective of the project FSA is to provide management with an independent assessment of the progress, quality, and attainment of project objectives, at defined milestones within the project based on company policies and procedures, also provide management with an assessment of the adequacy of project Functional Safety Management methodologies [3] and that the methodologies are applied consistently across all projects, then evaluate the internal controls of proposed Functional Safety processes at a point in the development cycle where enhancements can be easily implemented and processes adopted. Another objective is to provide management with an assessment of the adequacy of Functional Safety controls implemented [7] and provide management with an evaluation of the project's Functional Safety metrics.

The purpose of the Internal Functional Safety Assessment Procedure is to determine if project Functional Safety is on track or not, also to determine why and offer specific advice on how to get it back on track. In addition to determining whether what causes the issues are systemic and how to avoid them in the future.

The scope concerns the Functional Safety Requirements from ISO26262 including methods and guidelines, work products and requirements from the following ISO26262 chapters: Ch1 to Ch10.

This procedure is applicable to the complete automotive company or branch side and applied at ASILs (B), C, and D.

In relation to internal FS Assessments, we find the ISO 26262 requirements described in ISO 26262 Standard, Part 2 (section 6.4.9)

1.3.2.2 Project FSA procedure:

Performing FSA requires a competent staff and most often relies heavily on subjectivity, especially when applied to the latest phases of the security lifecycle. It covers who is responsible, when, and how they should be carried out as well as highlights the key information that is required as an input into each FSA stage and the expected outputs.

Overall Responsibility: Safety Manager.

Project FSA Responsible: Functional Safety Project FSA Responsible (selected from Functional Safety Team as appropriate).

Project FSA Initiation (when): for the chosen project at the end of the Concept, Development, and Validation & Testing Phases.

Project FSA Input(s)/Resources: access to project staff, documentation, and Work Products.

Project FSA Roles and Responsibilities:

- Safety Manager: Report to management the findings of the audit.

- Functional Safety Team: Plan, prepare and conduct the Project FSA. Prepare and issue the project FSA report. Follow-up on any nonconformance reports issued throughout the audit. Report to the Safety

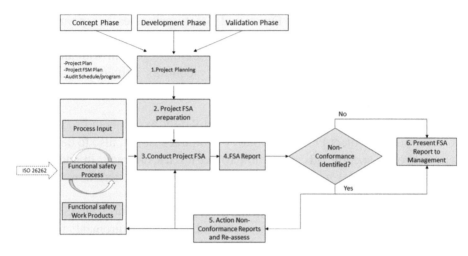

Figure 1.5 FSA steps.

Manager the findings of the audit. Escalate any non-conformance as necessary to the Safety Manager.

• Project Functional Safety Responsible: Provide and arrange access to project personnel and documentation. Action any non-conformance reports issued during the audit.

Project FSA Procedural Steps:

An overview of the project FSA procedure is shown in the diagram below. Project FSA procedure starts by Planning the Project FSA in accordance with the FSA Program Schedule, e.g. notify the Project Functional Safety Responsible that an assessment will be conducted., define the scope, objectives, criteria, etc. Include the Project Functional Safety Planning activities in the Project FSA Plan. The next step is to prepare for the Project FSA including the identification of the assessment team, level of independence from the project subject to assessment, review of any necessary documentation and work products, etc. The third step is to carry out the Project FSA, including interviews, reviews, etc. Then prepare the FSA Report including any non-conformances identified. After that the Project Functional Safety Responsible should address any identified non-conformances and the FSA Team conduct follow-up assessments as necessary. The findings of the FSA should be communicated to management as an output Report.

1.3.3 External or supplier functional safety audit and functional safety assessment

The External or Supplier functional safety audit assesses the adequacy of the documented functional safety processes and conformance of these processes to the ISO 26262 from an external or supplier organisation point of view. Figure 2 can be similarly applied to supplier related aspects.

Definitions, procedures, objectives and scopes are related to the external or supplier entity working with the automotive company. But in general, will follow and ensure compliance with ISO26262 and related safety processes.

1.4 Functional Safety Audit / Assessment Planning

To plan the Functional Safety Audit and/or FSA, the planning of the internal, external, or supplier audit and/or assessment should consider the following:

- Audit/Assessment Objectives - define what is to be achieved by the audit/assessment.

- Audit/Assessment Scope - could be complete systems audit / project assessment, covering the entire Functional Safety Management System, or it could be limited to one or more procedures specific to an activity or department.

- Pre-audit/Assessment Notice - notify those involved of the audit/assessment schedule, and scope.

- Systems audit/Assessment - the adequacy of the documented system is under review. To give confidence that the system is such that a satisfactory level of procedural control is present to ensure that the end product of the activity will meet the requirements of the company.

- Conformance audit - determine if the documented procedures are implemented correctly and are achieving the desired results.

- Audit Criteria - e.g. policies, procedures, regulations, management system standards, contract requirements, codes of conduct. The criteria are used to measure the level of conformance.

The output of the Functional Safety Audit / Assessment Planning phase is the Functional Safety Audit / Assessment Plan.

1.5 Functional Safety Audit / Assessment Preparation

When it comes to performing a Functional safety audit/Assessment, the preparation phase comes first. Starting by identifying a person or a team that will represent the Functional safety audit and/or FSA. Deliver any training needs to the Functional safety audit and/or FSA team to inform them about what is necessary for the functional safety audit/assessment and make them familiarized with the process, then Schedule planning for the Functional safety audit/Assessment, e.g. timings, documentation reviews, interviews preparation, preparation of checklists.

As a result, the preparation phase could include a document review (e.g. review of documented procedures, previous audit reports, work products, etc.) and a Functional Safety Audit/Assessment Checklist (tailored to the objectives of the audit/assessment).

1.6 Functional Safety Audit / Assessment Performance

After the preparation phase, we start performing the functional safety audit/ assessment following the planning steps and procedures to identify gaps and defects in the established ISO 26262 standard. Regularly performing a Functional Safety Audit is recommended as it reduces the likelihood that improper implementation of the process will impact other projects; inconsistencies arise later in the assessment.

1.7 Functional Safety Audit / Assessment Report and Follow-up

Following the functional safety assessment, the report including the results of the assessment (passed, conditionally accepted, or rejected) must be generated. The report might include a recommendation for conditional acceptance. In this case, the conditions of acceptance will also be detailed. If the recommendation in a functional safety assessment report is a rejection, appropriate corrective actions that might also be mentioned in the report should be planned and executed, and then the functional safety assessment should be repeated.

In the future, we will try to improve the functional safety process, especially in terms of time and reuse. The main challenge in today's automotive engineering is the efficiency of the evaluation and auditing processes.

1.8 Conclusion

In summary, a good plan is the key to making a functional safety audit and assessment succeed. When this is done before and during the development phase of a project, potential risks and gaps can be early detected and corrected. It can also help us to properly manage functional safety within the project by determining recommendations for corrective actions, as the audit and assessment are usually carried out by people with engineering and process experience associated with the ISO 26262 standard.

References

[1]   Birch J. et al. (2013) Safety Cases and Their Role in ISO 26262 Functional Safety Assessment. In: Bitsch F., Guiochet J., Kaâniche M. (eds) Computer Safety, Reliability, and Security. SAFECOMP 2013. Lecture Notes in Computer Science, vol 8153. Springer, Berlin, Heidelberg. https://doi.org/10.1007/978-3-642-40793-2_15

[2]   A. Nardi and A. Armato, "Functional safety methodologies for automotive applications," *2017 IEEE/ACM International Conference on Computer-Aided Design (ICCAD)*, 2017, pp. 970-975, doi: 10.1109/ICCAD.2017.8203886.

[3]   G. Xie, Y. Li, Y. Han, Y. Xie, G. Zeng, and R. Li, "Recent Advances and Future Trends for Automotive Functional Safety Design Methodologies," in *IEEE Transactions on Industrial Informatics*, vol. 16, no. 9, pp. 5629–5642, Sept. 2020, doi: 10.1109/TII.2020.2978889.

[4]   Y. Chang, L. Huang, H. Liu, C. Yang and C. Chiu, "Assessing automotive functional safety microprocessor with ISO 26262 hardware requirements," Technical chapters of 2014 International Symposium on VLSI Design, Automation, and Test, 2014, pp. 1–4, doi: 10.1109/VLSI-DAT.2014.6834876.

[5]   M. Safar, "Asil decomposition using SMT," 2017 Forum on Specification and Design Languages (FDL), 2017, pp. 1–6, doi: 10.1109/FDL.2017.8303902.

[6]   W. M. Goble and J. V. Bujkowski, "Extending IEC61508 reliability evaluation techniques to include common circuit designs used in industrial safety systems," Annual Reliability and Maintainability Symposium. 2001 Proceedings. International Symposium on Product Quality and Integrity (Cat. No.01CH37179), 2001, pp. 339–343, doi: 10.1109/RAMS.2001.902490.

[7] A. Munir, "Safety Assessment and Design of Dependable Cybercars: For today and the future.," in IEEE Consumer Electronics Magazine, vol. 6, no. 2, pp. 69–77, April 2017, doi: 10.1109/MCE.2016.2640738.

[8] ISO, "Iso 26262:2011 - road vehicles – functional safety," International Organization for Standardization in ISO 26262, 2011.

[9] ISO, "Iso 26262:2018 - road vehicles – functional safety," International Organization for Standardization in ISO 26262, 2018.

2

Comparison between AUTOSAR Platforms with Functional Safety for Automotive Software Architectures

Youssef El Kharaz[1], Saad Motahhir[2], Abdelaziz El Ghzizal[3]

[1]Technopolis Rabat shore B5 Sala El Jadida; youssef.elkharaz@usmba.ac.ma
[2]ENSA, USMBA; saad.motahhir@usmba.ac.ma
[3]EST, USMBA; abdelaziz.elghzizal@usmba.ac.ma

Abstract

In the next Vehicle generations, connected and highly developed driving cars will have an important impact on the networking architecture and the interconnection between ECUs(Electronic Control Unit). The automotive industry begins to develop new and efficient strategies to improve the performance of the global system. AUTOSAR organization as part of this industry tries to present plenary solutions especially software architectures for new technologies in this field. Thus, in this chapter, we present the aspects of new E/E architectures with upcoming technologies. We discuss a new solution presented by AUTOSAR organization to implement new software requirements for next-generation cars. This solution aims to provide a safe environment for the features that require complex data processing and to communicate with AUTOSAR and non AUTOSAR Platforms. We summarize a comprehensive comparison between AUTOSAR adaptive and AUTOSAR classic in terms of functionality and application area. We provide functional Safety preliminaries for the global E/E architectures.

2.1 Introduction

Recent cars like connected and autonomous vehicles becoming a state of art in the automotive industry. That leads directly to an increase in the percentage of electronic components [1] and embedded systems within an automobile. Moreover, automobiles are inherently safety critical. The uses of ECUs in modern day cars are growing exponentially, each for specific functionality. Present technologies such as infotainment [2], Car-to-X technology [3], and Autonomous cars [4]the automobile industry achieved remarkable mile-stones in manufacturing reliable, safe, and affordable vehicles. Because of significant recent advances in computation and communication technologies, autonomous cars are becoming a reality. Already autonomous car prototype models have covered millions of miles in test driving. Leading technical companies and car manufacturers have invested a staggering amount of resources in autonomous car technology, as they prepare for autonomous cars' full commercialization in the coming years. However, to achieve this goal, several technical and non-technical issues remain: software complexity, real-time data analytics, and testing and verification are among the greater technical challenges; and consumer stimulation, insurance management, and ethical/moral concerns rank high among the non-technical issues. Tackling these challenges requires thoughtful solutions that satisfy consumers, indus-try, and governmental requirements, regulations, and policies. Thus, here we present a comprehensive review of state-of-the-art results for autonomous car technology. We discuss current issues that hinder autonomous cars' develop-ment and deployment on a large scale. We also highlight autonomous car applications that will benefit consumers and many other sectors. Finally, to enable cost-effective, safe, and efficient autonomous cars, we discuss several challenges that must be addressed (and provide helpful suggestions for adop-tion require high level computing power. This means that the architecture network shall process and transport a large size of data in a short time.

To establish these new technologies in the software architecture besides the existing requirements including safety and security, a dynamic software environment is needed. In this context, the AUTOSAR organi-zation formed by OEMs and suppliers recognized that these new require-ments couldn't be implemented by the existing software architectures [5]. In general, AUTOSAR provides solutions for different software requirements throughout layered architecture that separates the software from the micro-controller where the reuse of software components for other applications is workable [6], as depicted in Figure 2.1. The AUTOSAR consortium provides two standards called Classic Platform and Adaptive Platform that are used for different goals and requirements. The classic platform is already established in the industry and intended to fulfill stringent real-time requirements with cost-optimized processors. On the other hand, AUTOSAR adaptive Platform was introduced recently to handle applications with outstanding performance

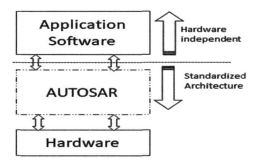

Figure 2.1 Application software separated from the hardware.

requirements such as highly automated driving. The combination of these trends will pave the way for the upcoming E/E architectures.

Throughout this chapter, we present the key aspects of heterogeneous platforms, and we highlight the AUTOSAR Adaptive platform standard with its last release. We study the main differences between classic AUTOSAR and Adaptive AUTOSAR. We describe how to establish the communication between Classic and Adaptive AUTOSAR, and finally, we provide the functional safety preliminaries to establish safely the automotive applications.

2.2 Overview of the Future E/E Architectures

2.2.1 Combination of different software platforms

Nowadays, networking architectures of new cars can be found in different domains, such as connectivity, infotainment, electrification, etc. Each domain has a specific ECU with a specific methodology of development. In addition of infotainment ECUs which are using Linux or other operating systems, and AUTOSAR Classic Platform is the standard for deeply embedded ECUs, another type of ECUs must arise with different characteristics that have to go along and interconnected with existing E/E architectures to respond to the demands from the new automotive applications.

2.2.2 Service oriented communication

The Communication between ECUs throughout signals using protocols like CAN, LIN, FlexRay and MOST fit very well for transmission data within an automobile. Meanwhile, advanced technologies like infotainment and Car-to-X demand higher bandwidth which are not highly supported by these protocols because of limited characteristics [5].

The service-oriented communication [7] is a flexible and efficient way to interconnect systems and their subscribers based on applications that provide services on the communication network. The future E/E architecture will be strictly based on the combination of the service-oriented paradigm and the existing communication protocols. For that AUTOSAR as a consortium formed by OEMs and their suppliers are in charge to standardize a new platform which is called AUTOSAR Adaptive platform using existing standards.

2.3 The AUTOSAR Adaptive Platform

The AUTOSAR Adaptive platform is a standardized architecture for high-performance ECUs to build safety systems such as highly driving and autonomous systems. Figure 2.2 depicts the global architecture of this platform.

Starting from the release 1.0.0 until the last release R20-11, several concepts affecting the Adaptive Platform have been introduced thereby adding new functionality to the platform, one of the core features of this adaptive platform is called AUTOSAR Runtime for Adaptive Applications (ARA). ARA gives users all the interfaces and infrastructure needed to communicate and execute adaptive applications into the system and allows data exchange between ECUs regardless of their internal architectures. In addition, this runtime offers direct access to the operating system functions known as the "Minimum Real time system Profile" (PSE51).

The module operating system interface based on a subset of POSIX [8] is responsible for run-time resource management such as signals, timer,

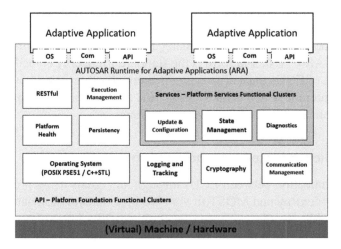

Figure 2.2   Adaptive AUTOSAR platform.

and thread handling for all adaptive applications and functional clusters that establish the platform.

In the AUTOSAR Adaptive platform, applications are not totally bounded by static scheduling and memory management but are free to allocate memory on their current need and break down their tasks thanks to object-oriented programming.

The Execution Manager module is an element of the architecture responsible for start-up and stopping the AUTOSAR Adaptive Applications, and responsible for providing the necessary resources during the execution period of the applications. To ensure the communication between local applications and applications on other ECUs including the interaction with the Adaptive platform services, middleware protocols must be defined. The most noticeable changes in the use of AUTOSAR Adaptive are the universal use of Ethernet based communication systems. For the release R20-11, new technology added to support the Ethernet protocol is related to 10BASE-T1S [9] which is specified by IEEE802.3cg.This new feature allows easy integration of devices into automotive Ethernet using the multidrop configuration. Furthermore, it is localized on layers 1 and 2 of the OSI model and is to be supported by Classic Platform as well as Adaptive.

AUTOSAR organization has an extensive release plan for adaptive AUTOSAR. The latest release date was in November 2020.The main focus for this release is to enhance the security and communication (10BASE-T1S, ara Communication Groups) by adding new functionality to the platform. Additionally, some concepts target the Classic and Adaptive Platform and reinforceing the interaction between the two platforms.

2.4 AUTOSAR Foundation

ECUs of future cars consist of different architecture platforms to offer required functionalities such as highly automated driving, connectivity, chassis, and infotainment, note that each system might be classified as a safety-critical or no safety-critical part. It's necessary a middleware communication between platforms achieve complete functionality of the global system. For that, the AUTOSAR organization defined a separate standard called AUTOSAR Foundation. The main goal of the foundation standard [10] is to enforce interoperability between the AUTOSAR platforms, and to ensure compatibility between:

- Classic- and Adaptive Platform.

- Non-AUTOSAR platforms to AUTOSAR platforms.

2.5 Classic AUTOSAR Vs Adaptive AUTOSAR

Classic AUTOSAR was the first achievement of a standardized software architecture for ECUs, then Adaptive AUTOSAR has just supplied a new architecture to meet new OEM requirements. In the classic platform, the application layer handles the communication between software components through the runtime environment (RTE) with the help of the OS. The RTE acts as an abstraction layer in between ECUs and establishes inter-ECU communication via a specific communication network. In the Adaptive platform, the applications utilize the "AUTOSAR Runtime for Adaptive Applications," also known as ARA.

This runtime environment gives users standardized interfaces to efficiently integrate different applications into the system. ARA offers mechanisms for ECU-internal and inter-network communications as well as access to basic services such as diagnostics and network management. The adaptive AUTOSAR applications are formed in software components that

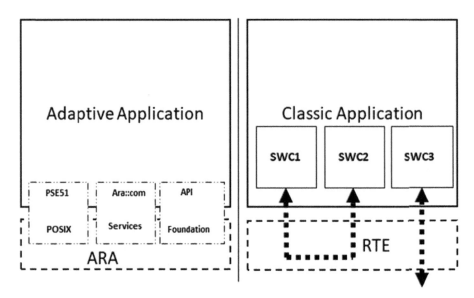

Figure 2.3 Difference between applications within adaptive AUTOSAR and classic AUTOSAR.

communicate via services. These services may be requested or provided. In addition, the application programmer can directly access a subset of operating system functions. In terms of communication, the AUTOSAR Adaptive defines a new feature called ara::com. ara::com is a standard C++ API based on SOA more specifically based on SOME/IP. In parallel rough, AUTOSAR classic covers the communication between software components using RTE signals.

The operating system presents some of the key differences between the adaptive AUTOSAR platform and the classic AUTOSAR platform. In adaptive AUTOSAR applications are no longer tied to very strict and static scheduling and memory management but are certainly free to create and destroy tasks and allocate memory as needed, including the use of C++ as a programming language in contrast to C in classic platform.

An important point in AUTOSAR platforms is the update and configuration management. The Adaptive platform now offers the option of removing, updating, or adding individual applications, while the Classic platform can only replace the entire ECU code during an update.

Another property of the Adaptive platform is its transition to an exclusively service-oriented architecture paradigm, which offers greater flexibility in system design. Applications provide their functionality as a service via the

```
Void Function_Runnable(){

Uint8 Data_Input;
Uint8 Data_Output;

RTE_ReadSignal_Input(&Data_Input);

/* brief Implementation*/

RTE_WriteSignal_Output(&Data_Output);
}
```

```
Class FuncionServiceProxy{
public:
        /* brief implementation of FuncionServiceProxy
class*/
Explicit FuncionServiceProxy(HandleType &handl);

        /* brief public member for the BrakeEvent*/
events::BrakeEvent Brake vent;

        /* brief Public Field for UpdateRate*/
fields::UpdateRate UpdateRate;

        /* brief public member for the Adjust method*/
methods::Calibrate calibrate;

        /* brief public member for the Adjust method*/
methods::Adjust Adjust ;

};
```

Figure 2.4   An example illustrates the code implementation of RTE and ARA::COM.

Table 2.1   Comparison between AUTOSAR adaptive and AUTOSAR classic.

	AUTOSAR Classic	**AUTOSAR Adaptive**
Application Interface	Use RTE with the help of OSEK	Use ARA
Operating system	OSEK	POSIK
Programming language	C	C++
Remove/update the application	Remove or update the entire ECU	Remove, add, or update individual application
Communication protocols	Signal based communication network bus (CAN, LIN)	Service oriented communication based on ethernet over (SOME/IP)
Utilization	Implement deeply embedded systems	Implement high performance applications like high-automated driving
Functional safety	Up to ASIL D	Up to ASIL D

Adaptive platform, and they can use the services that are offered. In the other hand the focus of the Classic platform is primarily on signal-oriented communication. Nonetheless, it is also possible to use AUTOSAR Classic in a service- based way for communication between multiple ECUs. In practice, the main properties of the AUTOSAR Adaptive and Classic platforms complement one another, it may therefore be assumed that ECUs based on both standards will be used in future vehicles resulting in a heterogeneous architecture.

2.6 Communication Between AUTOSAR Platforms

How to communicate between Adaptive and Classic ECUs is a mandatory question. In such a scenario, the ECUs which are interconnected over ethernet use service-oriented communication over SOME/IP. In this example, the AUTOSAR Classic ECU1 is connected to multiple bus systems to which other ECUs are connected (Figure 2.5).

ECU1 operates as a gateway in this configuration, and it is responsible for transferring the message signals from the bus side into a service so that they can be accessed directly by the AUTOSAR Adaptive platform. The communications layout is a fixed component of the design for AUTOSAR ECUs, whether it is a Classic or Adaptive platform, Because the configuration format is different for the two platforms, it is necessary to map the service configuration in the form of a conversion. The situation is somewhat more multifaceted for communicating with an AUTOSAR Classic ECU whose operation is exclusively signal based. In this scenario, the ECU1 is designed as a signal

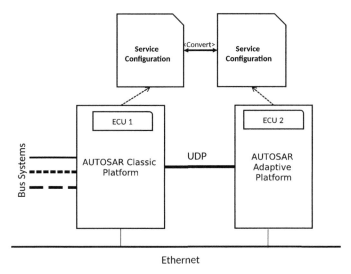

Figure 2.5   Inter-platform communication over Ethernet and SOME/IP.

gateway, and it converts message signals directly into UDP frames [11] on ethernet then the AUTOSAR Adaptive ECU converts signals from the UDP frame to a service that is available within ECU2.

2.7 Safety Preliminaries for E/E Architectures

2.7.1 Functional safety overview

In the Automotive Industry, we prefer to define safety as freedom from hazardous situations that can cause damage, physical or material to humans in the car and on the road.

During the process of realizing the Automotive applications, guaranteeing safety is always a prerequisite. According to the definition in ISO 26262, functional safety means no unpredictable faults due to hazards risks by the malfunctioning behavior of E/E Architecture [12].The standard ISO 26262 has two versions, the first which has been introduced in 2011, and the recent one in December 2018 [13].

The AUTOSAR platforms should consider the Functional safety characteristics from the beginning of the development phase, as it may impact the system and software architectural design decisions. Thus, the AUTOSAR Adaptive or classic specifications include requirements related to functional safety. Aspects such as the complexity of the system design can be relevant for the achievement of functional safety in the automotive industry.

Table 2.2   ASIL determination in ISO 26262.

Severity	Exposure	Controllability		
		C1	**C2**	**C3**
S1	E1	QM	QM	QM
	E2	QM	QM	QM
	E3	QM	QM	A
	E4	QM	A	B
S2	E1	QM	QM	QM
	E2	QM	QM	A
	E3	QM	A	B
	E4	A	B	C
S3	E1	QM	QM	A/QM
	E2	QM	A	B
	E3	A	B	C
	E4	B	C	D

2.7.2 ASIL determination

In ISO 26262, Automotive Safety Integrity Level (ASIL) is a risk classification scheme that helps the developers to meet the functional safety requirements for safety automotive applications. The ASIL is based on Hazard Analysis and Risk Assessment through estimating the severity, exposure, and controllability of each hardware/software component behavior. ISO 26262 identified four ASILs Levels A, B, C, and D plus Quality Management (QM) which is not considered as an ASIL, while ASIL D represents the highest degree of automotive hazard and ASIL A represents the lowest degree.

To determine an ASIL we need a combination of severity, exposure, and controllability. Exposure indicates the probability of causing harm, severity Indicates the extent of harm, and controllability indicates the ability to avoid the harm through the timely reactions of the drivers [14]the automotive functional safety design has been challenged by multiple factors, such as the complexity of the new generation automotive Electrical and Electronic (E/E. Table-II illustrate the ASIL Determination based on severity, exposure, and controllability.

2.8 Conclusion

Currently, Autonomous, and connected car technology are developing and presenting more challenges especially in terms of safety. In the software part, AUTOSAR as an organization provides solutions that fit very well for these challenges.

Adaptive AUTOSAR is not designed to replace Classic AUTOSAR in terms of functionality, but we need them both to coexist and cooperate to meet the needs and challenges of the automotive industry. To determine which one to use in a specific application, we need to start by answering the following questions: How much computing power do we need? What are our time requirements? And how dynamic is our application?

In the coming works we are going to implement both AUTOSAR platforms separately in a target and perform a real time comparison.

References

[1] J. G. Kassakian et D. J. Perreault, « The future of electronics in automobiles », in *Proceedings of the 13th International Symposium on Power Semiconductor Devices & ICs. IPSD '01 (IEEE Cat. No.01CH37216)*, Osaka, Japan, 2001, p. 15–19. doi: 10.1109/ISPSD.2001.934550.

[2] D.-K. Choi, J.-H. Jung, S.-J. koh, J.-I. Kim, et J. Park, « In-Vehicle Infotainment Management System in Internet-of-Things Networks », in *2019 International Conference on Information Networking (ICOIN)*, Kuala Lumpur, Malaysia, janv. 2019, p. 88–92. doi: 10.1109/ICOIN.2019.8718192.

[3] J. Schafer et D. Klein, « Implementing Situation Awareness for Car-to-X Applications Using Domain Specific Languages », in *2013 IEEE 77th Vehicular Technology Conference (VTC Spring)*, Dresden, Germany, juin 2013, p. 1–5. doi: 10.1109/VTCSpring.2013.6692589.

[4] R. Hussain et S. Zeadally, « Autonomous Cars: Research Results, Issues, and Future Challenges », *IEEE Commun. Surv. Tutor.*, vol. 21, n° 2, p. 1275–1313, 2019, doi: 10.1109/COMST.2018.2869360.

[5] « Furst and Bechter - 2016 - AUTOSAR for Connected and Autonomous Vehicles The.pdf ».

[6] S. Bunzel, « AUTOSAR – the Standardized Software Architecture », *Inform.-Spektrum*, vol. 34, n° 1, p. 79–83, févr. 2011, doi: 10.1007/s00287-010-0506-7.

[7] G. L. Gopu, K. V. Kavitha, et J. Joy, « Service Oriented Architecture based connectivity of automotive ECUs », in *2016 International Conference on Circuit, Power and Computing Technologies (ICCPCT)*, Nagercoil, India, mars 2016, p. 1–4. doi: 10.1109/ICCPCT.2016.7530358.

[8] V. Atlidakis, J. Andrus, R. Geambasu, D. Mitropoulos, et J. Nieh, « POSIX abstractions in modern operating systems: the old, the new, and the missing », in *Proceedings of the Eleventh European Conference*

on *Computer Systems*, London United Kingdom, avr. 2016, p. 1–17. doi: 10.1145/2901318.2901350.

[9]   G. Huszak et H. Morita, « On the 10BASE-T1S preamble for multi-drop », in *2019 Global Information Infrastructure and Networking Symposium (GIIS)*, Paris, France, déc. 2019, p. 1–6. doi: 10.1109/GIIS48668.2019.9044963.

[10] « Foundation Release Overview », p. 26.

[11] « Specification of UDP Network Management », p. 103.

[12] « ISO - Standards », *ISO*. https://www.iso.org/standards.html (consulté le 26 juillet 2021).

[13] « ISO 26262-1:2018(en), Road vehicles — Functional safety — Part 1: Vocabulary ». https://www.iso.org/obp/ui/#iso:std:iso:26262:-1:ed-2:v1:en (consulté le 27 juillet 2021).

[14] G. Xie, Y. Li, Y. Han, Y. Xie, G. Zeng, et R. Li, « Recent Advances and Future Trends for Automotive Functional Safety Design Methodologies », *IEEE Trans. Ind. Inform.*, vol. 16, n° 9, p. 5629–5642, sept. 2020, doi: 10.1109/TII.2020.2978889.

3

Hardware-in-the-Loop System for Electronic Control Unit Software and Calibration Development

M. Elhag[1], M. Najoui[2], A. Jbari[3]

[1]Powertrain Engineering Department, AVL Maroc SARL AU, 11100, Sala Al Jadida, Morocco
[1,2,3]E2SN Team, ENSAM-Rabat, Mohammed V University in Rabat, Morocco
[1]mahdi.elhag@avl.com; [2]m.najoui@um5r.ac.ma; [3]atman.jbari@ensam.um5.ac.ma

Abstract

In today's world, the powertrain system complexity has increased drastically, whereas the development time for powertrain hardware and software is still being decreased due to the cost reduction strategies of automotive companies. This chapter will offer an overview of Electronic Control Unit architectures and their usage in the automotive domain. To highlight the real implementation constraints, we report a cost-effective solution based on a hardware-in-the-loop (HIL) system that was developed to perform real-time closed-loop simulations for testing the engine hardware components and the software in the ECU, as well as the calibration of these software functions. The proposed implementation method uses cheaper components compared to the existing systems in academia or industry. Also, it concentrates on representing a new combination of software and hardware tools to simulate the engine and signals. By utilizing this system, it is expected to have fast, robust, and cost-effective software and calibration development processes for the automotive industry.

3.1 Introduction

In the development cycle of engine controls, several challenges are encountered in the determination of control targets, controller design, and software

33

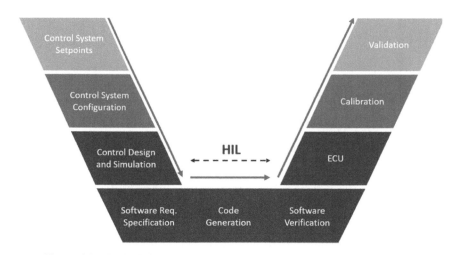

Figure 3.1   Typical V-cycle for automotive controllers' design and validation.

implementation [1]. The design and validation of such a system are performed according to the V-cycle presented in Figure 3.1. Nowadays, automotive control and calibration strategies are much more complicated compared to the past the car industry. Thus, engineers must put more effort into developing and calibrating systems within a short period and satisfying strict demands from customers and emissions regulations [2]. Likewise, there is a need for advanced hardware-software interdependent systems to handle the challenges in the design. After the software verification step using software-in-the-loop (SIL), the ECU hardware and calibration are tested in HIL to finalize the cycle. In the past, several researchers tackled the topic of producing a mobile solution for the HIL to be used in software and calibration development [3, 4, 5, 6]. However, there have been many shortcomings in the solutions provided in the literature, especially in terms of the hardware for generating signals and real-time communication between the device under test and the plant model environment [7, 8, 9].

As an alternative solution, a system called Automotive Systems Simulation HIL (A2S-HIL) is planned to be utilized in several calibration and software development applications such as training activities, engine model validation, ECU function tests, engine/after-treatment, and component calibration activities. Likewise, this tool enhances the process of studying software functions in the ECU, the closed-loop system structures, software-hardware interactions in automotive systems, internal combustion engines, and training activities. Also, software validation tests can be performed for ECU functions on this system as an open-loop system. Similarly, any engine model that runs in an open- loop system can be validated via A2S-HIL as well. To close the control loop, the system requires outputs from the engine plant model and feedback signals from ECU to the model. In case the validation tests resulted in a mature engine model, calibration tasks that require open-loop or closed-loop tests on a dynamometer can also be handled using an on-desk A2S-HIL. According to several studies, mature models can have an error of less than 31.4% [10]. For that, the accuracy of any engine model can be determined by calculating the error in the operation points with the highest time spent. Even though the engine model cannot represent all the transient cases as well as the real-life, the pre-calibration tasks can still be done on A2S-HIL. Especially, open-loop functions can easily be calibrated without the need for feedback coming from the ECU. For example, catalyst efficiency monitoring does not require any signal to be returned to the engine model. Consequently, the following calibration tasks are proper to the open-loop approach: on-board diagnostics (OBD) component monitoring calibration, OBD release condition calibration, model-based function calibration, environmental trip precondition, etc.

In this work, we suggest avoiding expensive hardware devices while ensuring the same accuracy. Therefore, the engine model validation with a real ECU for control and calibration applications is performed on the same system without major modifications. Comparably, the postposed system can run several driving cycles in open-loop and closed-loop. Also, most of the hardware connections are replaced with software communication structures. This allowed maintaining all the system components synchronous in almost real-time. In parallel, the concept of functional mock-up (FMU) is introduced into engine HIL simulations. Besides that, the structure of the system was established on the Matlab®/Simulink® platform including all the subsystems such as cam/crank signal generator based on the cycle inputs, AVL Cruise™ M engine model, a calibration data acquisition software, and AVL Concerto™ was used for plotting.

This chapter is organized as follows: Section 2 presents an introduction to the automotive electronic controls, Section 3 defines the method and implementation details of each subsystem, and Section 4 shows the results of the system in open and closed-loop cycles. Finally, Section 5 concludes the chapter. Moreover, the remainder of this chapter elaborates on the creation of a testing environment, beginning with the importance of such test systems to the automotive sector and an example application on a real engine control unit.

3.2 Automotive Embedded Systems Overview

3.2.1 Automotive systems

Automotive systems engineering is the study of the engineering methods implementation in the area of vehicle technology. Therefore, the vehicle is targeted as the system [11]. From that point of view, a typical vehicle mostly consists of an engine, a fuel system, an ignition system, an electrical system, an exhaust system, a powertrain, a suspension, steering system, a brake system, a frame, and a body.

Mainly for an internal combustion engine, the constraints of design include:

- Obtaining the maximum performance,

- Reducing the fuel consumption,

- Reducing tailpipe emissions,

- Ensuring the safety of passengers,

- Protecting the components of the vehicle.

Consequently, several control algorithms are applied to ensure injecting the optimal fuel into the cylinders. For such systems, there are sensors and actuators to be controlled electronically.

3.2.2 Electronic control units

The electronic control unit (ECU) is a complex embedded system that manages all engine functions from the driver inputs to the exhaust gas reduction systems. It consists of a microcontroller, filters, and transistors placed in a single printed circuit board (PCB).

Nowadays, over 40 ECUs exist in a normal car and even more than 150 in a luxurious cars. ECUs are used in a variety of systems in a vehicle like ADAS, infotainment, body control, and comfort system among others. This means that there is a huge amount of data that is being processed by an ECU during each ride. ECUs are utilized in various car categories ranging from light vehicles to heavy vehicles including tractors.

Eventually, the main features of ECUs usage in cars are:

- Systematic transfer of data,

- Dependability and security,

- Efficient data network,

- Diagnostics,

- Assistance in real-time decision making,

- Improved quality of service.

The ECUs are classified based on their functions such as engine control module (ECM), body control module (BCM), electronic brake control module (EBCM), powertrain control module (PCM), transmission control module (TCM), suspension control module (SCM), door control unit (DCM), battery management system (BMS), and many other. All the functions of automotive, from basic window movements to a critical safety or fuel injection function, are being controlled by ECUs. They collect, analyze the data, and decide the actions based on defined parameters.

For example, the engine control module (ECM), also named engine control unit (ECU), ensures that the car operates at optimal results. It reads most of the sensors in the engine in order to manage the air-fuel mixture and regulate the emission. To successfully control an engine, the microcontroller inside the ECU read the accelerator, brake, and clutch pedal signals to calculate the mass of fuel to be injected. Then it gives the opening signal to the injectors at the optimal time [12]. The ECM controls four main parts of any car: air-fuel ratio, idle speed, variable valve timing, and ignition timing. In terms of the air-fuel ratio, the ECM uses Lambda sensors to regulate the oxygen to fuel ratio in the exhaust. That is to detect an engine status as rich or lean. For the idle speed, the ECM counts on a crankshaft position sensor that tracks the engine revolutions. The variable valve timing system controls the valves opening and closing to either increase power or fuel economy. Lastly, the ECM controls the ignition spark timing. Accurate control of this provides more power and fuel economy.

As another example, the battery management system (BMS) is mostly seen in hybrid and electric vehicles. Mainly, lithium-ion batteries, are a pack of multiple cells that need to be monitored closely because failure in a single cell may affect the performance of the whole battery. Besides that, it checks various parameters of the battery such as the health of the battery, state of charge, the current account, voltage, and power to optimize the performance. BMS also supports the recharging of the batteries from the regenerative braking.

Though all ECUs function independently, communication between each other is needed as well. This communication is mostly via the CAN bus and managed by the body control module (BCM). BCMs fall under the category of ECUs, but as a gateway to connect the other ECUs. They consist of processors that manage multiple body functions in a vehicle such as windows, lights, and wipers.

Apparently, ECUs are designed to support a specific number of inputs and outputs. In addition to that, multi-purpose controllers exist nowadays. Mostly, Tier 1 suppliers such as Bosch, Continental, and Delphi usually use the same ECU hardware with several software releases. Considering this high level of complexity, the automotive engineers follow the development cycle of an ECU from requirements definition until the validation of all functions. Apparently, tests can go to different levels to assess the system.

3.2.3 Hardware-in-the-loop systems

A Hardware-in-the-Loop (HIL) simulation is a technique that is used in the development and test of complex real-time embedded systems. It is a method where real signals from a controller are connected to a system that simulates dynamics, allowing the controller to consider it is in the full product. Normally, test and design iteration take place as though the physical system is being used. It became easier to run through thousands of possible test cases to properly verify the control algorithm in a low cost and brief time associated with actual real-life tests. Therefore, it adds the complexity of the plant to the test platform. Moreover, the Device-Under Test (DUT) is the ECU that its functions need to be developed, tested, or calibrated.

Such systems are designed to achieve the following objectives:

- HIL systems shorten the development time of software and calibration,

- Testing engine behavior before or after deploying to real life applications,

- Debugging the system outputs for certain scenarios according to dangerous cases.

A typical HIL test system consists of three primary components: a real-time environment, I/O interface devices, and a graphical user interface. The real-time computer is the heart of the HIL because it provides deterministic execution of most of the HIL components such as hardware I/O management, data logging, input signal generation, and scenario execution. A real-time operating system is typically necessary to obtain an accurate simulation of the plant models that are not physically present in the test. The I/O interfaces are analogue to digital, digital to analogue, and CAN signals that interact with the tested controller. The graphical interface handles the communication with the real-time processor to give commands and visualize results. Mostly, this component also provides configuration management, test automation, analysis, and reporting tasks.

3.3 HIL for Engine Calibration: A Case Study

3.3.1 System setup

The implementation of the HIL system requires robust signal generation and communication between all components. The basic obstacles to having

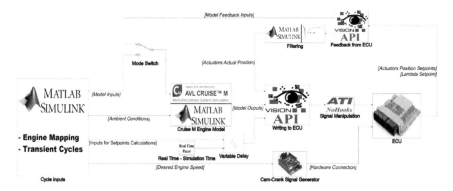

Figure 3.2 HIL overview and components used for system realization.

a robust A2S-HIL system are Cam/crank signal generation for steady and transient engine speed simulation, the delivery of the model outputs to the ECU, and, transmitting feedback signals to the engine model (in closed-loop mode).

To power up the ECU, a 3 A 12 V power supply was used. Also, a push-button switch was used in the electrical circuit to simulate the ignition key of the engine. As for the accelerator pedal of the A2S-HIL system, there are two modes for accelerator pedal position (APP) control: the cycle input mode and the external mode. In the cycle input mode, the APP was fed from the engine model to ECU by writing its electrical signal value to the memory directly. While, in the external mode, a potentiometer was used to control the pedal position by generating the analogue signal for the APP pin.

In Figure 3.2, the subsystem "Cruise™ M Engine Model" represents the plant model which is a 1-liter spark-ignition engine. It receives the desired ambient pressure and temperature of test execution from the block "Cycle Inputs" which also delivers the desired engine speed signal to the "Cam-Crank Signal Generator" board. The plant can function in an open or closed loop according to the status of "Mode Switch". Subsequently, the outputs of the model are transferred to the ECU through the "Signal Manipulation" strategy that writes the electrical values of the physical quantities to the memory of the controller. In closed-loop mode, the actuators and lambda setpoints are filtered to obtain the actual feedback values. The throttle and the waste-gate valve signals are sent to the engine.

The cam/crank signals serve as the core of all ECU calculations. Depending on the software that the ECU uses, the profile of these signals may vary. The crankshaft position sensor can be considered the primary sensor for electronic fuel injection and ignition. It produces a pulse for each tooth on

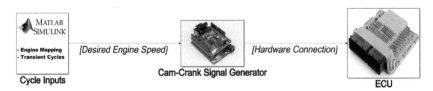

Cycle Inputs

Cam-Crank Signal Generator

ECU

Figure 3.3   Signal flow for the generated camshaft and crankshaft signals from Simulink® to ECU through the cam-crank signal generator board.

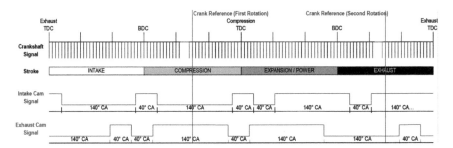

Figure 3.4   Typical crankshaft and camshaft signals for 1-liter spark-ignition engine [14] (TDC: Top Dead Centre, BDC: Bottom Dead Centre, CA: Crank Angle.)

the crankshaft which gives the engine speed and the shaft's instantaneous position [13]. In A2S-HIL, the frequency of the crankshaft signal was varied according to the desired engine speed. A board containing an Arduino Uno and a Sparkfun® CAN-Shield was utilized to generate the crankshaft signal as shown in Figure 3.3. This "Cam-Crank Signal Generator" board establishes real-time CAN communication between itself and Simulink®. Furthermore, intake and exhaust camshaft position sensor signals were also necessary to measure the camshaft positions for the valves' opening and closing. These signals were generated in full synchronization with the crankshaft signal as shown in Figure 3.4.

For instance, the required PWM frequency to be generated by the microcontroller for each sample was estimated by converting the maximum desired engine speed (4000 revolutions/minute = 66/6 revolutions/second) in the HIL system into sampling time according to the number of degrees in the crankshaft. Since each revolution of the crankshaft contains 720 degrees (1 degree = 1 cycle), then the maximum frequency of the "Cam-Crank Signal Generator" microcontroller is expressed in Equation 3.1.

$$PWM[hz] = \frac{66[rev]}{6[sec]} = \frac{66*720[cycle]}{6[sec]} = 7920[hz] \qquad (3.1)$$

3.3.2 Signal manipulation

The second obstacle in the system structure was handling the signal transfer from the engine model in Simulink to the integrated calibration software which managed the memory of the ECU. The engine model outputs such as air mass flow, throttle upstream pressure, intake manifold temperature, and pressure were converted into utilizable values using each sensor's transfer curve. To ensure the real-time operation of the system, the reading and the writing sequences of the engine model outputs back and forth to the ECU channels, were synchronized using ATI® software [15].

As illustrated in Figure 3.5, all the required sensor values were written to the ECU in the same way. Mass airflow (MAF), manifold air pressure (MAP), the throttle upstream pressure, and pedal position sensors' signals had a 100 ms sampling time for writing from Simulink to the ECU. Moreover, the manifold air temperature and the throttle upstream temperature signals had a 1000 ms sampling time since the temperature signals have slow dynamics. As for reading the inputs of the engine model, ATI® Vision® was used with a 100 ms sampling time.

Some inversed transfer curves were used on the model outputs to calculate the desired voltage values. Lambda and the throttle valve position setpoints were fed back to the engine model in the closed-loop simulation as presented in Figure 3.6. Environmental parameters were obtained from the driving cycle data such as engine mapping, transient set, New European Driving Cycle (NEDC), Worldwide harmonized Light vehicles Test Cycle (WLTC), Real Driving Emission (RDE), etc.

A2S-HIL needs to operate on a PC capable of running Simulink®. However, the regular operating systems used are not real-time operating systems (RTOS). For that, another solution was proposed to approach real-time functionality as much as possible. A variable delay is applied to all signals during each simulation step which equals the difference between the real-time and simulation time. Such a Matlab® algorithm ensures that the plant model in Simulink® and the cam-crank generator run in 100 ms sampling

Figure 3.5   Signals flow from cycle inputs to ECU by ATI® Vision® and NoHooks®.

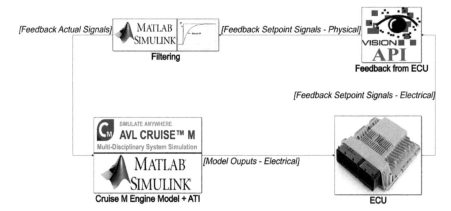

Figure 3.6 Signals flow from cycle inputs to ECU by ATI® Vision® and NoHooks®.

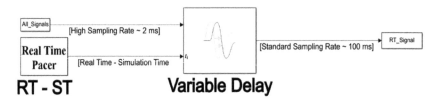

Figure 3.7 The approach of slowing down signals flow to ensure real-time running.

time. Actively, the engine model was slowed down to be closer to real-time as shown in Figure 3.7.

3.3.3 Engine model

AVL Cruise™ M is a simulation platform used for modelling physical systems for development and testing purposes [16]. The engine model used in the proposed A2S-HIL was created in Cruise™ M and then exported to Simulink.

As in Figure 3.8, the plant model requires the following inputs: engine speed, the start of injection angle, injection time, fuel rail pressure, engine coolant temperature, intake valve opening/closing angle (IVO/IVC), exhaust valve opening/closing angle (EVO/EVC), ignition angle, ambient pressure/temperature (P_ATM/T_ATM), throttle valve position, wastegate position, and lambda setpoint. On the other side, the engine model generates the following outputs: MAF (MF_A), air filter downstream pressure/temperature (P_11/T_11), turbocharger downstream pressure/temperature (P_21/T_21), throttle upstream pressure/temperature (P_2_1/T_2_1), intake manifold air pressure/temperature (P_IM/T_IM), and throttle upstream pressure/

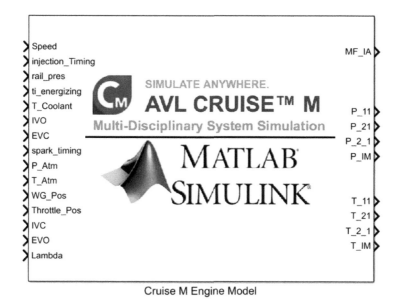

Figure 3.8   The inputs and outputs of the engine plant model in Simulink.

temperature (P_2_1/T_2_1). The other outputs such as exhaust mass flow, brake mean effective pressure, brake specific fuel consumption, and exhaust manifold pressure/temperature were not fed to the ECU because they weren't essential for the first version purposes of the suggested A2S-HIL. Thereafter, the ECU uses those mandatory sensor measurements to calculate the setpoints for lambda and actuators' positions such as throttle valve. In Simulink®, the engine model runs faster than in real-time. To overcome this issue and be close as much as possible to real-time, some delays were added to slow down the simulation.

3.3.4 Overall system

To be used for different applications, the proposed A2S-HIL can operate in two main functional modes. Firstly, the open-loop mode is useful to evaluate the plant model accuracy and execute software functional testing in ECU as presented in Figure 3.9. The inputs of the engine are given from real testbench data to obtain the same outputs on a specific test case. For example, to calibrate voltage-pressure conversion curves, the system runs in an open-loop several times to evaluate the impact of the proposed calibration on other functions.

Secondly, the closed-loop mode in which all the plant model inputs except the environmental conditions and engine coolant temperature are fed

Figure 3.9   Simplified system components for open-loop mode overview.

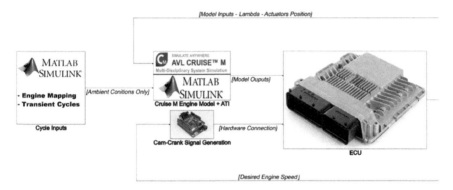

Figure 3.10   Simplified system components for closed-loop mode overview.

back from ECU to the model as illustrated in Figure 3.10. This mode supports the calibration of actuators like the throttle valve by running the system and observing the set points and actual values according to different controller adjustments.

3.4 Experimental Evaluation and Measurements

3.4.1 Experimental setup

For HIL systems, there are sensors to measure the inputs such as crankshaft position, ambient conditions, throttle position, and richness. Similarly, the outputs of the system refer to signals that participate in the control algorithm. Contrarily, the actuators such as the throttle must be taken into consideration [17]. To reduce effort and increase the efficiency of A2S-HIL, a physical throttle was wired to the ECU as shown in Figure 3.11. With such an upgrade, the setpoint value of the actuator was fed into the controller, and the actual value was obtained from the throttle position sensor directly. A similar concept can be applied to the injectors, spark plugs, and fuel rail pressure. Nevertheless, the cost and objective of the setup must be taken into consideration.

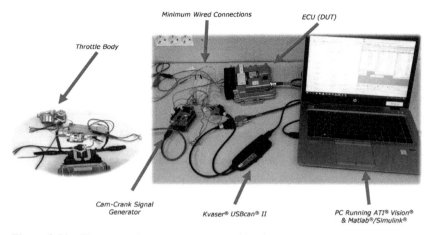

Minimum Wired Connections   ECU (DUT)

Throttle Body

Cam-Crank Signal Generator   Kvaser® USBcan® II   PC Running ATI® Vision® & Matlab®/Simulink®

Figure 3.11   The system layout shows the used hardware and overall occupied space.

In Figure 3.12, the device under-test (DUT) was an opened ECU. Such controllers allow reaching all the memory in real-time through CAN. The Kvaser® USBcan® II has two channels. One of them communicated the DUT to the PC, and the other passed the engine speed frequency to the "Cam-Crank Signal Generator" via CAN as well. Moreover, the wired connection consisted of the power supply, crank signal, intake/exhaust cam signals, and the ignition key switch. Whereas in the left side of the figure, there were two cables representing the throttle positive and negative position sensor signals.

As highlighted before, the pins of the other sensors were bypassed from within the memory of the ECU. For example, the MAF sensor electrical signal address in the RAM was replaced by the engine model value of Matlab®/Simulink®. This strategy has reduced the number of wires and digital-analogue converters (DAC) required to simulate the whole system.

The steady and transient tests were done on a real MPV (Multi-Purpose Vehicle). A laptop with ATI® Vision® was used to record the input acquisitions via Kvaser® Leaf Light® v2 CAN device as shown in Figure 4.2. To save cost, the same Kvaser® USBcan® II can be utilized on the vehicle by connecting a single CAN channel to the ECU. As with other alternatives, Vector® CAN devices are compatible with ATI® Vision®. In case other tools will be used, the datasheets of products must be checked.

3.4.2 Experiment measurements

The A2S-HIL overall evaluation was done on the subsystem level in both open-loop and closed-loop modes. A steady engine mapping and a transient

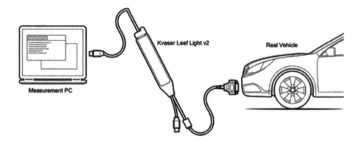

Figure 3.12   Connection layout between vehicle and laptop.

real driving emission (RDE) cycle were run for open-loop validation for the engine mode as well as the cam-crank signal generator board. A transient cycle with slow dynamics and an external accelerator pedal position signal was tested for the closed-loop mode validation.

3.4.3 Engine model accuracy results

Being a mandatory component of A2S-HIL, the engine model was developed by AVL especially for this ECU. Testing its accuracy was done by validation cycles.

Based on the validation process that AVL follows for engine model validation, the output of the engine model was compared with the measured cycle outputs in various cycles. Since the aim of this activity was just to validate the model itself, an open-loop mode of A2S-HIL was utilized. Eventually, the errors in all the model signals were calculated according to the following equation [18]:

$$\text{Error}\,(\%) = \frac{\text{Signal}_{\text{Real}} - \text{Signal}_{\text{Model}}}{\text{Signal}_{\text{Real}}} * 100\% \qquad (3.2)$$

Therefore, the average errors were 7.71% and 15.01% in mass air flow and manifold air pressure outputs respectively as shown in Figures 3.13 and 3.14. The figures of model accuracy show the real measurement mapped data on the upper right side, the model simulation result on the upper left side, the difference between measurement and simulation results on the lower left side, and the error distribution according to y = x line to show displacement of simulation results from real engine data on the lower right side. All outputs were mapped according to engine speed (x-axis) and fuel quantity (y-axis). Concretely, the lower left side plots showed the blue and red regions were the sections with the highest error Additionally, the plot of the lower

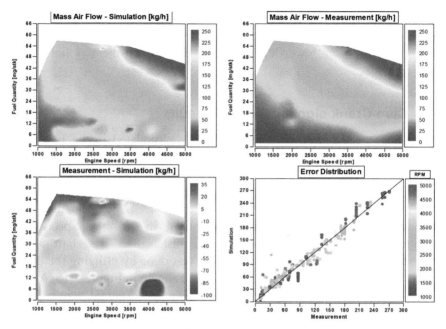

Figure 3.13 AVL Cruise™ M engine model MAF accuracy results.

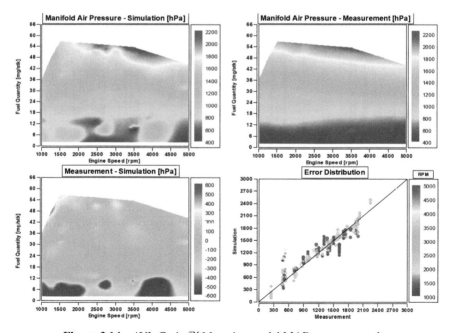

Figure 3.14 AVL Cruise™ M engine model MAP accuracy results.

right side illustrated the deviation of the model from the real engine and the engine speed value for each point. Optimally, a 100% accurate model will have all green on the lower left side and all points on the x=y line on the lower right side.

3.4.4 Cam-Crank signal generator results

The engine speed obtained from A2S-HIL was updated in the ECU with a 100 ms sampling time. Based on the capacity of the microcontroller, the input engine speed values were converted to frequencies during signal generation. Thus, the proper profile of the cam/crank signals could be maintained whereas the frequency of the signal, sent to the ECU, was changed. This approach can respond to even the fast transitions in the engine speed signal while maintaining full synchronization of camshaft and crankshaft signals. Therefore, the validation of this subsystem was done by utilizing the data in Figures 3.15 and 3.16, the obtained results are shown in Figure 3.15. A cycle was executed for 200 seconds, and the engine speed signal was recorded from the HIL system. Then it was compared to the real engine speed of the vehicle. Assuredly, the "Cam-Crank Signal Generator" subsystem achieved a maximum 0.04% error according to Equation 3.2.

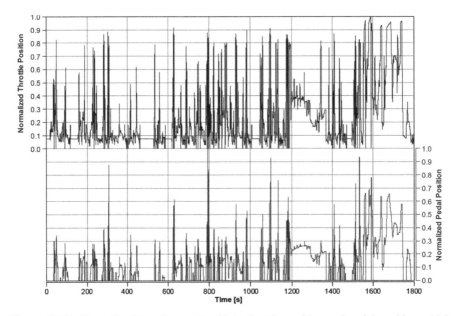

Figure 3.15 Normalized transient real vehicle throttle position and pedal position which were used in analyze the transient results.

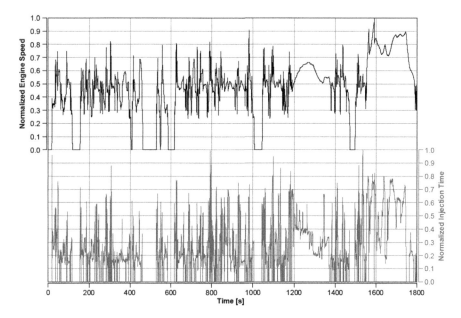

Figure 3.16 Normalized transient real vehicle engine speed and injection time which were used in analyze the transient results.

The other cycle inputs such as the accelerator pedal position and the environmental conditions were converted to electrical signals and fed to the ECU in 100 ms sampling time as presented in Figure 3.17. Accordingly, the real vehicle accelerator pedal signal (blue line) was almost equal to the HIL system results (red line) the "Signal Manipulation" subsystem processed the data transfer with a maximum error of 0.02%. This minor error value was due to the time spent by the engine speed signal to pass from Simulink® to the "Cam-Crank Signal Generator" board.

3.4.5 Open-loop performance

A2S-HIL open-loop was implemented to do OBD calibrations, software testing, and engine model accuracy testing. For most of the OBD calibration tasks, the closed-loop system is not necessary [19]. Thus, the engine model outputs, and the cycle data were written to the ECU directly. For engine model accuracy testing, the cycle data were given from a pre-recorded test as inputs to the model and the modeled sensor output values were compared to the actual values in Figures 3.18 and 3.19. Accordingly, an RDE cycle was run in A2S-HIL. The cycle inputs were given to the engine model and the output signals were written to the ECU's memory. In ATI® Vision®, a

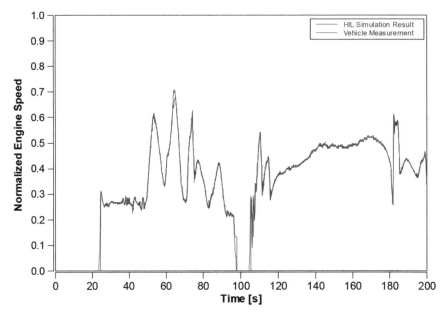

Figure 3.17   A transient cycle comparison of the engine speed from a vehicle measurement and the engine speed generated by the microcontroller.

record containing the engine model's output was taken and compared with the actual measurements of the cycle. The manifold air pressure and mass airflow signals were validated with the real engine data. The minor errors observed at the time range 120–130 and 180–190 seconds were the same blues and red regions in Figures 3.13 and 3.14. These results validated the "Signal Manipulation" subsystem as well.

3.4.6 Closed-loop performance

As mentioned, the open-loop mode was designed for engine diagnostics calibration activities. Contrarily, the closed loop is important for actuators controller calibration applications only. The accuracy of the engine model affected the performance of A2S-HIL. In this setup, the accelerator pedal position (APP) was manipulated externally using a potentiometer under a constant engine speed. All actuators were responding to the change of APP by targeting different setpoints. The engine model had received the setpoints and generated outputs that moved the actual sensor measurements to the set-points. The results of simulating a driver pedal signal input starting from 0 to full press are shown in Figure 3.20.

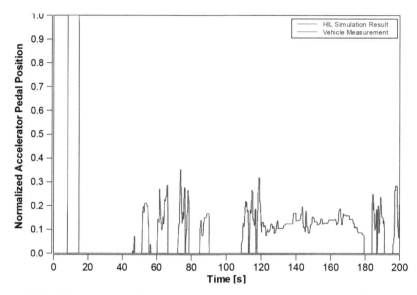

Figure 3.18 The comparison between the accelerator pedal position signal from a vehicle measurement and A2S-HIL.

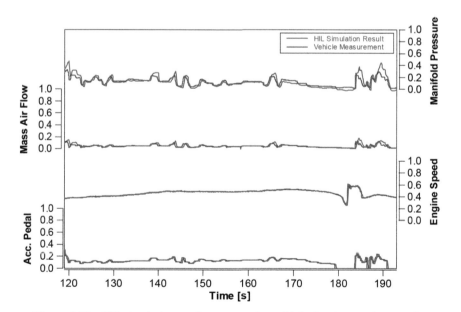

Figure 3.19 HIL simulation results compared to vehicle data in open-loop mode.

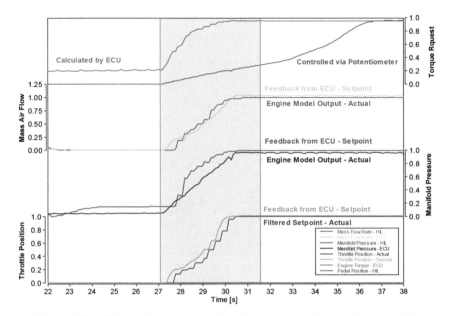

Figure 3.20   HIL results in a steady closed-loop cycle with an accelerator pedal.

3.5 Conclusion

The use of hardware-in-the-loop systems setup for engine calibration tasks was challenged by cost, time, and manpower to achieve targets. In this chapter, the A2S-HIL working principle and subsystems were discussed. The first step of this study targeted cost-effectiveness in HIL systems as well as the high sampling rate. The challenges faced such as the generation of cam and crank signals, modeling the engine in Cruise™ M, and the integration of all components was a feasibility study for the A2S-HIL concept. The areas of application such as software development, engine controls, OBD, and calibration were introduced briefly. Concretely, the Cruise™ M engine model validation for calibration purposes was done using a real ECU and read measurements. Also, the sampling time of all the system components was 100 ms in open-loop and closed-loop simulations. Despite the robustness of signals and integration of components, there were deficiencies to be addressed.

In the upcoming research, an engine model with higher accuracy will be introduced. Moreover, several OBD calibration tasks use cases will be performed. Similarly, actuator models such as throttle valves will replace the position setpoints. Likewise, an after-treatment model of this engine model will be developed to perform calibration on the exhaust path. Additionally, the sampling time for the overall system will be improved. Finally, future works will focus as well on implementing this cost-effective HIL testing methodology on a hybrid control unit.

3.6 Acknowledgements

The authors gratefully acknowledge all the affiliates of AVL List GmbH for providing access to the software and hardware used in this project. Also, everyone who was involved in developing this challenging project from the family of AVL is appreciated. The authors would like to thank Accurate Technologies Inc. (ATI) for their suggestions and support.

References

[1] Shahbakhti M., Li J., Hedrick J. (2012). "Early model-based verification of automotive control system implementation," in Proceedings of American Control Conference (ACC), doi:10.1109/ACC.2012.6314852

[2] Lee, S. Andert, J. Neumann D., Querel C. et al. (2018). "Hardware-in-the-Loop-Based Virtual Calibration Approach to Meet Real Driving Emissions Requirements," in Proceedings of SAE Int. J. Engines 11(6):1479–1504, https://doi.org/10.4271/2018-01-0869.

[3] Pedersen M.M., Hansen M.R., Ballebye M. (2011). "A Cost-Effective Approach to Hardware-in-the-loop Simulation," in Proceedings of Jabłoński R., Březina T. (eds) Mechatronics. Springer, Berlin, Heidelberg

[4] Isermann R., Schaffnit J., Sinsel S. (1999). "Hardware-in-the-loop simulation for the design and testing of engine-control systems," in Proceedings of Control Engineering Practice, ISSN: 0967-0661, Vol: 7, Issue: 5, Page: 643–653

[5] Kendalla I. R., Jones R. P. (1999). "An investigation into the use of hardware-in-the-loop simulation testing for automotive electronic control systems," in Proceedings of Control Engineering Practice, ISSN: 0967-0661, Vol: 7, Issue: 11, Page: 1343–1356

[6] Jaikamal V. (2009). "Model-based ECU development – An Integrated MiL-SiL-HiL Approach," in Proceedings of SAE World Congress & Exhibition. SAE Technical Paper 2009-01-0153, https://doi.org/10.4271/2009-01-0153

[7] Nanjundaswamy H., Tatur, M. Tomazic, D. Dahodwala, M. et al. (2011). "Development and Calibration of On-Board-Diagnostic Strategies Using a Micro-HiL Approach," in Proceedings of SAE

World Congress & Exhibit. SAE Technical Paper 2011-01-0703, doi:10.4271/2011-01-0703

[8] Nabi S., Balike M., Allen J., Rzemien K. (2004). "An Overview of Hardware-in-the-Loop Testing Systems at Visteon," in Proceedings of SAE Technical Paper 2004-01-1240, https://doi.org/10.4271/2004-01-1240.

[9] Zheng, L. (2019). "Research and Implementation of Automated HIL Simulation Test Platform for ECU of Automotive Engine," in Proceedings of International Journal of Mechatronics & Applied Mechanics, (6), 45–56.

[10] Wong K.I., Wong P.K., Cheung C.S. (2015). "Modelling and Prediction of Diesel Engine Performance using Relevance Vector Machine," in Proceedings of International Journal of Green Energy, 12:3, 265–271, DOI: 10.1080/15435075.2014.891513

[11] Maurer, M., Winner, H., eds. (2013). Automotive Systems Engineering. Springer, 3–15, Heidelberg New York Dordrecht London. Doi: 10.1007/978-3-642-36455-6

[12] Hui D., Bo H., Dafang W., Guifan Z. (2011) "The ECU Control of Diesel Engine Based on CAN," in Proceedings of 2011 Fourth International Conference on Intelligent Computation Technology and Automation, pp. 734–736, doi: 10.1109/ICICTA.2011.191.

[13] Nwagboso C.O. (1993). "Sensors and systems for crankshaft position measurement," in Proceedings of Nwagboso C.O. (eds) Automotive Sensory Systems. Springer, Dordrecht

[14] Drosescu R. (2017). "Virtual engine management simulator for educational purposes," in Proceedings of Materials Science and Engineering, 252, 012099 doi:10.1088/1757-899X/252/1/012099

[15] Rogers, D., Church, M., Patel, U., and Menon, C. (2013). "The Evolution of Rapid Prototyping," in proceedings of Symposium on International Automotive Technology, SAE Technical Paper 2013-26-0082

[16] Nagi M., Iorga D., Cărăbaş I., Irimescu A., Laza I. (2011). "Simulation of a Passenger Car Performance and Emissions Using the AVL-Cruise™ Software," in Proceedings of the Politehnica University of Timisoara, Faculty of Mechanics, Termotehnica, Romania

[17] Zhou B., Yang J., Xi G., Chen P. (2013). "Research on a hardware-in-the-loop simulation platform of engine sensor- fault," in Proceedings of Zhongguo Jixie Gongcheng (China Mechanical Engineering), 24(9), 1181–1185. DOI: 10.3969/j.issn.1004-132X.2013.09.011,

[18] De Nola F., Giardiello G., Molteni A., Picariello R. (2018). "Enhancing the accuracy of engine calibration through a computer-aided calibration

algorithm," in Proceedings of Energy Procedia, Volume 148, Pages 916–923, ISSN 1876-6102, doi: 10.1016/j.egypro.2018.08.094

[19] Ranal R.A., Nascimento R.A. (2011). "HIL Simulation System for Application of Electrical Diagnosis and OBD in Diesel Engines ECUs," in Proceedings of Virtual Powertrain Conferences, Speakers Information. Direct Link: https://issuu.com/ranal/docs/hil

SECTION 2

Utilizing Embedded Systems for UAVs

4

Processor in the Loop Experiments of an Adaptive Trajectory Tracking Control for Quadrotor UAVs

Hamid Hassani[1]*, Anass Mansouri[2], Ali Ahaitouf[1]

[1]SIGER Laboratory, Faculty of Sciences and Technology, Sidi Mohammed Ben Abdellah University, Fez, Morocco
[2]SIGER Laboratory, School of Applied Sciences, Sidi Mohammed Ben Abdellah University, Fez, Morocco
*Email: hamid.hassani@usmba.ac.ma

Abstract

Quadrotor drones are highly maneuverable rotary wing vehicles, which are vulnerable to modeling uncertainties, state variable couplings, and aerodynamic perturbations. These factors pose an issue that warrants a robust controller. In this chapter, we explore the design of an adaptive sliding mode controller for trajectory tracking of an uncertain quadrotor. The suggested approach delivers good tracking performance, despite the severe impact of uncertainties and disturbances. The Lyapunov theory is used to analyze the closed-loop stability of the entire system and to calculate the adaptive laws. The efficacy of the suggested controller is evaluated under the influence of disturbances and modeling inaccuracies. Moreover, Processor-in-the-Loop (PIL) experiments on an STM32F429 discovery board are carried out to confirm the workability of the suggested method.

4.1 Introduction

During the past few years, quadrotor drones have experienced a boost. Thanks to the new upcoming technologies, quadrotors have become the preferred solution to cover multiple tasks including surveillance, package delivery, mapping, cinematography, etc [1], [2]. The autonomy in the quadrotor

59

system is achieved via position and attitude control. However, designing an effective flight control system is challenging because the vehicle dynamic is under actuated, complex, and highly nonlinear [3]. A further challenge is to stabilize the aircraft under some atmospheric disturbances such as wind gusts [4]–[6].

Recently, many advanced control approaches have been implemented for autonomous vehicles to achieve high performance in terms of accuracy and robustness [7]–[10]. In [11], a sliding mode controller (SMC) was designed based on neural networks to steer the vertical and rotational motions of a quadrotor aircraft under the influence of external perturbations. However, the control of lateral and longitudinal motions was not considered in this research. In [12], an adaptive SMC tuned using a particle swarm optimization algorithm is developed for the path following of a quadrotor under uncertainties of mass and matrix of inertia. In [13], an efficient technique for solving the problem of under-actuation is designed for a quadrotor aircraft; in addition, a compounded control scheme is employed to steer the vehicle in the 3-dimensional space. However, the influence of disturbance/ uncertainties was not considered in this work. The authors in [14], developed a finite-time control system for the position and orientation of the quadrotor. In addition, a self-tuning methodology is developed to approximate the unknown bound of exogenous disturbances. Aiming at mitigating the chattering, Muñoz et al. [15] compared the efficiency of three forms of the super twisting algorithm applied for the tracking control of the quadrotor altitude. The idea developed in [16], treats the robust path following using a nonlinear control scheme that incorporates the backstepping strategy with the integral SMC. In addition, the super twisting algorithm is also utilized to mitigate the chattering and enforce the robustness against the additive perturbations. In order to vanish the impact of external time-varying/sustained disturbances, a control law based on a nonlinear technique with finite-time convergence property is synthetized for a quadrotor subjected to unmodeled dynamics [17]. Mofid et al. [18] developed an adaptive super-twisting terminal SMC for a quadrotor UAVs in the existence of input-delay, uncertainties, and external disturbances. The validity of this method has been demonstrated through simulation and real outdoor flight. In [19], a backstepping SMC strategy combined with an active disturbance rejection control was proposed to improve the tracking accuracy of a quadrotor in the presence of model uncertainties and wind disturbances. A method to control the quadrotor orientation in aggressive maneuvers is treated in [20], where a disturbance observer is used to estimate external perturbations and unknown Coriolis terms. Besnard et al. [21] proposed a SMC working together with a sliding mode observer to raise the performance of the control system in path tracking missions.

Based on the above cited articles, in this chapter, we synthesize a simple adaptive nonlinear control law for path following of an underactuated quadrotor exposed to time-varying disturbances. The suggested controller is employed for the position and orientation to improve the tracking ability despite the severe impact of wind disturbances. The newly designed control algorithm efficiently rejects the impact of uncertainties and disturbances. Processor in the loop implementations was carried out to confirm the feasibility of our method. The performance of the suggested controller is examined under the influence of modelling uncertainties and wind disturbances.

The rest of this chapter is constructed as; the mathematical model of the aerial robot is investigated in section 2. Next, the proposed control scheme is detailed in section 3. PIL tests are highlighted in section 4. Section 5 concludes the chapter.

4.2 Quadrotor Modeling

Quadrotor drones are small multirotor lifted by four motors mounted at the endpoint of a symmetrical frame as schematized in Figure 4.1. In order to navigate to the desired position, the motors are forced to rotate at a specific speed, which makes the vehicle realizes the adequate maneuver needed to reach the desired location.

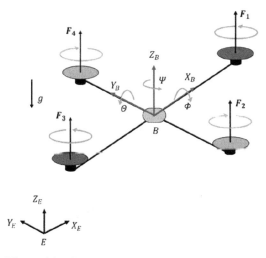

Figure 4.1 Configuration of the quadrotor aircraft.

Using the Newton-Euler method, the full dynamical model of the aerial robot is given by[22]–[25]:

$$
\begin{cases}
\ddot{\Phi} = \dot{\Theta}\dot{\Psi}\left(\dfrac{J_y - J_z}{J_x}\right) - \dfrac{J_r}{J_x}\dot{\Theta}\Omega + \dfrac{U_\Phi}{J_x} - \dfrac{K_\Phi}{J_x}\dot{\Phi}^2 + d_\Phi \\[2mm]
\ddot{\Theta} = \dot{\Phi}\dot{\Psi}\left(\dfrac{J_z - J_x}{J_y}\right) + \dfrac{J_r}{J_y}\dot{\Phi}\Omega + \dfrac{U_\Theta}{J_y} - \dfrac{K_\Theta}{J_y}\dot{\Theta}^2 + d_\Theta \\[2mm]
\ddot{\Psi} = \dot{\Phi}\dot{\Theta}\left(\dfrac{J_x - J_y}{J_z}\right) + \dfrac{U_\Psi}{J_z} - \dfrac{K_\Psi}{J_z}\dot{\Psi}^2 + d_\Psi \\[2mm]
\ddot{x} = \dfrac{U_1}{m}(\cos\Phi\sin\Theta\cos\Psi + \sin\Phi\sin\Psi) - \dfrac{K_x}{m}\dot{x} + d_x \\[2mm]
\ddot{y} = \dfrac{U_1}{m}(\cos\Phi\sin\Theta\sin\Psi - \sin\Phi\cos\Psi) - \dfrac{K_y}{m}\dot{y} + d_y \\[2mm]
\ddot{z} = \dfrac{U_1}{m}(\cos\Phi\cos\Theta) - g - \dfrac{K_z}{m}\dot{z} + d_z
\end{cases}
\tag{4.1}
$$

The aircraft's absolute position is denoted by $\chi = (x, y, z)$, while the attitude is described by Euler angles $\gamma = (\Phi, \Theta, \Psi)$. $J = diag\left(J_x, J_y, J_z\right)$ symbolizes the matrix of inertia. J_r denotes the motor inertia. K_i, $i = (\Phi, \Theta, \Psi, x, y, z)$ signify the aerodynamic friction constants. $(U_1, U_\Phi, U_\Theta, U_\Psi)$ are the four control inputs. d_h denotes the bounded additive perturbations.

The model of the quadrotor aircraft (Equation 4.1) can be described by a second order nonlinear system which has the following form:

$$
\dot{\chi} = A(\chi) + B(\chi)\mathcal{U} + d
\tag{4.2}
$$

With $\chi = \{\chi_h,\ h = 1,2,\dots,12\}$ stands for the quadrotor state variables given by $\{\Phi, \dot{\Phi}, \Theta, \dot{\Theta}, \Psi, \dot{\Psi}, x, \dot{x}, y, \dot{y}, z, \dot{z}\}$, $A(\chi)$ and $B(\chi)$ are given in Equation (4.3). $\mathcal{U} = [U_1, \tau_\Phi, \tau_\Theta, \tau_\Psi]^T$ represents the control signals. $|d_h| < \xi_h$, ξ_h is the upper limit of additive perturbations.

$$A(\chi) = \begin{bmatrix} \chi_2 \\ a_{1\Phi}\chi_4\,\chi_6 + a_{2\Phi}\chi_4 + a_{3\Phi}\chi_2^2 \\ \chi_4 \\ a_{1\Theta}\chi_2\chi_6 + a_{2\Theta}\chi_2 + a_{3\Theta}\chi_3^2 \\ \chi_6 \\ a_{1\Psi}\chi_2\,\chi_4 + a_{2\Psi}\chi_6^2 \\ \chi_8 \\ a_x\chi_8 \\ \chi_{10} \\ a_y\chi_{10} \\ \chi_{12} \\ -g + a_z\chi_{12} \end{bmatrix} ; \quad B(\chi) = \begin{bmatrix} 0 & 0 & 0 & 0 \\ B_\Phi & 0 & 0 & 0 \\ 0 & 0 & 0 & 0 \\ 0 & B_\Phi & 0 & 0 \\ 0 & 0 & 0 & 0 \\ 0 & 0 & B_\Psi & 0 \\ 0 & 0 & 0 & 0 \\ 0 & 0 & 0 & \dfrac{v_x}{m} \\ 0 & 0 & 0 & 0 \\ 0 & 0 & 0 & \dfrac{v_y}{m} \\ 0 & 0 & 0 & 0 \\ 0 & 0 & 0 & \dfrac{C_{\chi_1}C_{\chi_3}}{m} \end{bmatrix}$$

$$(4.3)$$

Where, $a_{1\Phi} = \dfrac{(J_y - J_z)}{J_x}, a_{2\Phi} = -\dfrac{J_r}{J_x}\Omega, a_{3\Phi} = -\dfrac{K_\Phi}{J_x}, a_{1\Theta} = \dfrac{(J_z - J_x)}{J_y}, a_{2\Theta} = \dfrac{J_r}{J_y}\Omega,$

$a_{3\Theta} = -\dfrac{K_\Theta}{J_y}, a_{1\Psi} = \dfrac{(J_x - J_y)}{J_z}, a_{2\Psi} = -\dfrac{K_\Psi}{J_z}, \qquad a_x = -\dfrac{K_x}{m}, \qquad a_y = -\dfrac{K_y}{m},$

$a_z = -\dfrac{K_z}{m}, \; B_\Phi = \dfrac{1}{J_x}, \; B_\Theta = \dfrac{1}{J_x}, \; B_\Psi = \dfrac{1}{J_z} .$

The quadrotor vehicle is underactuated, to circumvent this issue, virtual control signals are introduced as,

$$\begin{cases} v_x = \dfrac{U_1}{m}\left(C_{\chi_1}S_{\chi_3}C_{\chi_5} + S_{\chi_1}S_{\chi_5}\right) \\ v_y = \dfrac{U_1}{m}\left(C_{\chi_1}S_{\chi_3}S_{\chi_5} - S_{\chi_1}C_{\chi_5}\right) \end{cases}$$

$$(4.4)$$

Thereafter, the reference Euler angles are obtained as:

$$
\begin{cases}
\Theta_d = \arctan\left(\dfrac{v_x C_{\Psi_d} + v_y S_{\Psi_d}}{v_z} \right) \\[4mm]
\Phi_d = \arctan\left(C_{\Theta_d}\left(\dfrac{v_x S_{\Psi_d} - v_y C_{\Psi_d}}{v_z} \right) \right)
\end{cases}
\tag{4.5}
$$

With $v_z = \left(C_{\chi_1} C_{\chi_3} \right) \dfrac{U_1}{m}$

4.3 Controller Design

The flowchart of the suggested method is vividly illustrated in Figure 4.2. As seen, a hierarchical control methodology is used to steer the behaviors of the aircraft, where a position controller is developed to track the planned cartesian position and generate the desired Euler angles. Based on the position controller's output, an attitude controller is used to stabilize the orientation and ensure the achievement of the desired location. In order to deal with the negative impact of perturbations and reduce the tracking errors, both controllers are synthesized based on the SMC combined with an adaptive tuning methodology.

The aim of the proposed flight control system is to reduce the deviation between the desired state $\hbar_d = \{\Phi_d, \Theta_d, \Psi_d, x_d, y_d, z_d\}$ and the actual state $\hbar = \{\Phi, \Theta, \Psi, x, y, z\}$. Introduce the position and attitude tracking errors as follows:

$$
\begin{cases}
e_h = h - h_d \\
\dot{e}_h = \dot{h} - \dot{h}_d \\
\ddot{e}_h = \ddot{h} - \ddot{h}_d = A(h) + B(h)\mathcal{U} + d_h - \ddot{h}_d
\end{cases}
\tag{4.6}
$$

The control goal is to ensure that the tracking errors vanish asymptotically. To attain this objective, the sliding manifolds σ_h for the position and attitude subsystem are selected as:

$$
\sigma_h = \dot{e}_h + \lambda_h e_h
\tag{4.7}
$$

Where λ_h is a positive constant.

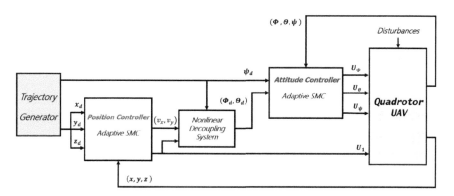

Figure 4.2 Proposed control scheme for the quadrotor aircraft.

The derivatives of the sliding manifolds are calculated as:

$$\dot{\sigma}_h = \ddot{e}_h + \lambda_h \dot{e}_h$$
$$= \left(\ddot{h} - \ddot{h}_d\right) + \lambda_h \dot{e}_h$$
$$= A(h) + B(h)\mathcal{U}_h + d_h - \ddot{h}_d + \lambda_h \dot{e}_h \tag{4.8}$$

In order to counteract the impact of unknown perturbations, the adaptive compensating reaching law is selected as,

$$\dot{\sigma}_h = -T_{1h}\sigma_h - T_{2h}\left|\sigma_h\right|^{\alpha_h} \, sign(\sigma_h) - \hat{\xi}_h sign(\sigma_h) \tag{4.9}$$

$$\dot{\hat{\xi}}_h = \mu_h\left|\sigma_h\right| \tag{4.10}$$

Where $\hat{\xi}_h$ is the estimate of $\xi_h \cdot T_{1h}, T_{2h}$, and μ_h are positive constants.
Using Equation (4.8) and Equation (4.9), the corresponding control signals for the attitude and position are obtained as:

$$U_h = \frac{1}{B(h)}(\ddot{h}_d - A(h) - \lambda_h \dot{e}_h - T_{1h}\sigma_h - T_{2h}\left|\sigma_h\right|^{\alpha_h} \, sign(\sigma_h) - \hat{\xi}_h sign(\sigma_h)) \tag{4.11}$$

Theorem 3.1: For the investigated quadrotor system defined by Equation (4.1), if the control signal is designed as Equation (4.11); then the entire system is asymptotically stable.

Proof. To demonstrate the above theorem, consider the following Lyapunov function:

$$V = \frac{1}{2}\left(\sigma_h^2 + \frac{\breve{\xi}_h^2}{\mu_h}\right)$$

(4.12)

Where $\breve{\xi}_h = \xi_h - \hat{\xi}_h$ is the error of adaptation.

The time-derivative of is given by:

$$\dot{V} = \sigma_h \dot{\sigma}_h - \frac{\breve{\xi}_h \dot{\hat{\xi}}_h}{\mu_h}$$

$$= \sigma_h \left(A(h) + B(h)\mathcal{U}_h + d_h - \ddot{h}_d + \lambda_h \dot{e}_h\right) - \frac{\breve{\xi}_h \dot{\hat{\xi}}_h}{\mu_h}$$

$$= \sigma_h \left(d_h - \mathcal{T}_{1h}\sigma_h - \mathcal{T}_{2h}|\sigma_h|^{\alpha_h} sign(\sigma_h) - \hat{\xi}_h sign(\sigma_h)\right) - \frac{\breve{\xi}_h \dot{\hat{\xi}}_h}{\mu_h}$$

$$= -\mathcal{T}_{1h}\sigma_h^2 - \mathcal{T}_{2h}|\sigma_h|^{\alpha_h+1} + d_h\sigma_h - \left(\xi_h - \breve{\xi}_h\right)|\sigma_h| - \frac{\breve{\xi}_h \dot{\hat{\xi}}_h}{\mu_h}$$

$$\leq -\mathcal{T}_{1h}\sigma_h^2 - \mathcal{T}_{2h}|\sigma_h|^{\alpha_h+1} + |d_h||\sigma_h| - \xi_h|\sigma_h| + \breve{\xi}_h|\sigma_h| - \frac{\breve{\xi}_h \dot{\hat{\xi}}_h}{\mu_h}$$

$$\leq -\mathcal{T}_{1h}\sigma_h^2 - \mathcal{T}_{2h}|\sigma_h|^{\alpha_h+1} + \left(|d_h| - \xi_h\right)|\sigma_h| + \breve{\xi}_h\left(|\sigma_h| - \frac{\dot{\hat{\xi}}_h}{\mu_h}\right)$$

(4.13)

Using equation (10), $\dot{V}$ becomes:

$$\dot{V} = -\mathcal{T}_{1h}\sigma_h^2 - \mathcal{T}_{2h}|\sigma_h|^{\alpha_h+1} \leq 0$$

(4.14)

As the Lyapunov function time derivative $\dot{V}$ is negative; then, the sliding surface σ_h and its first time-derivative are forced to converge to the origin. Thus, the system valued can achieve a stable state.

4.4 Processor-in-the-Loop Experiments

In this part, PIL experiments are performed to examine the controller's effectiveness and robustness against external unknown disturbances/uncertainties.

Table 4.1 Physical parameter of the quadrotor [26].

Parameters	Value	Parameters	Value
m (kg)	0.74	$K_{x,y,z}$ (N/m/s)	$5.567e-5$
$I_x = I_y (Kg.m^2)$	0.004	$K_{\Phi,\Theta,\Psi}$ (N/m/s)	$5.567e-5$
$I_z (Kg.m^2)$	0.0084	b (N·s^2)	$2.9e-5$
$l(m)$	0.21	d (N.m.s^2)	$1.1e-6$
g (ms^{-2})	9.81	J_r ($Kg.m^2$)	$2.838e-5$

Table 4.2 Parameter's value of the proposed controller.

Parameters	Value	Parameters	Value
$\lambda_{x,y,z}$	3	$\lambda_{\Phi,\Theta}$	17
$T_{1x,y}$	10	λ_{Ψ}	7
T_{1z}	3	$T_{1\Phi,\Theta,\Psi}$	30
$T_{2x,y,z}$	5	$T_{2\Phi,\Theta,\Psi}$	7
$\alpha_{\Phi,\Theta,\Psi,x,y,z}$	0.6	$\mu_{\Phi,\Theta,\Psi,x,y,z}$	2

An uncertain quadrotor model (Equation 4.1) is considered in the flight test and the corresponding parameters are grouped in Table 4.1. Besides, the value of the controller's parameters used in the PIL tests are listed in Table 4.2

Remark 1: PIL implementation is a powerful tool for verifying the efficiency of a control scheme on a hardware board, while the studied model is software[27]–[30]. Figure 4.3 illustrates the procedure used for the implementation of the proposed flight control algorithm. Firstly, the perturbed model of the quadrotor system (Equation 1) as well as the suggested control law (Equation 11) are programmed in MATLAB/Simulink software, this step is referred to as Model-in-the-Loop (MIL) where the controller as well as the plant are both modeled using Simulink blocks and tested in simulation on the host PC. Next, the code of the proposed controller is generated then integrated with the quadrotor model and tested again on simulation, this test is called Software-in-the-Loop (SIL). At the end, a MATLAB toolbox named *"Embedded Coder Support Package for STMicroelectronics Discovery Boards"* [31] is used to implement the controller on the STM32F429 discovery board. This toolbox enables the automatic generation of C++ code directly from MATLAB/Simulink, as well as the deployment of that code to the target board. During the PIL test, the generated code will be executed on the hardware while the quadrotor model is simulated on the host PC. The communication between the PC and the STM32F429 board is ensured via a USB wire.

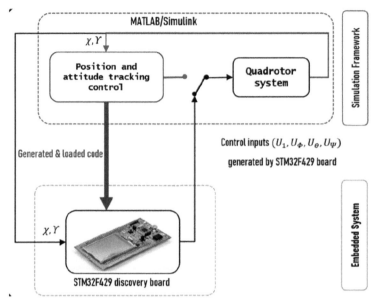

Figure 4.3   Procedure of the PIL experiment using the STM32F429 board.

In this scenario, the aircraft is commanded to follow a time-varying 3D-trajectory. Besides, the controller performance is evaluated under the influence of wind perturbations as well as $\mp 30\%$ of uncertainty in the rotary inertia. The disturbances caused by wind gusts are considered as $. 0.5\cos\left(\dfrac{\pi}{3}t\right)$

The results of PIL tests are illustrated in Figure 4.4 to 4.7. As it is shown in Figure 4.4, the quadrotor under the proposed controller can fulfill sufficiently good robustness as well as a reasonably precise tracking performance despite the impact of uncertainties. Moreover, the vehicle state achieves its target values rapidly and maintains a stable behavior during the flight mission. Figure 4.5 depicts the tracking performance of the quadrotor attitude. Clearly, the generated roll and pitch angles are well stabilized. Moreover, the vehicle's heading angle accurately tracked the time-varying reference angle. Figure 4.6 highlights the control signals generated for the three cases (without uncertainty, +30% of uncertainty and -30% of uncertainty). Obviously, these signals are reasonable and without chattering.

In order to clearly show the path following accuracy of the suggested controller, Figure 4.7 depicts the 3-dimensional space trajectories under the

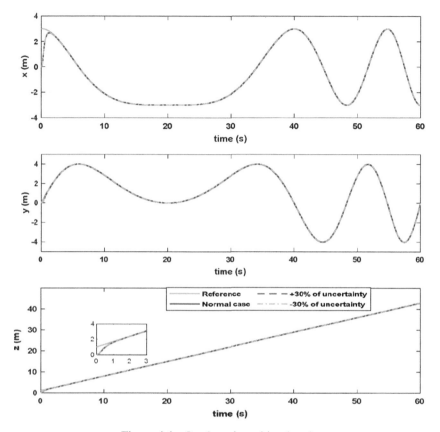

Figure 4.4   Quadrotor's position (x, y, z).

influence of modelling inaccuracies and wind disturbances. Obviously, the controller allows the quadrotor good tracking of the prescribed path.

Finally, it has been marked from the presented results that the position and orientation of the quadrotor subject to modeling uncertainties and external disturbances have good tracking performance. Moreover, the implementation of the suggested method on a real hardware confirms the results and demonstrations exhibited in the previous section.

4.5 Conclusion

A simple adaptive sliding mode position/attitude control for an unmanned quadrotor has been elaborated with consideration of modelling inaccuracies

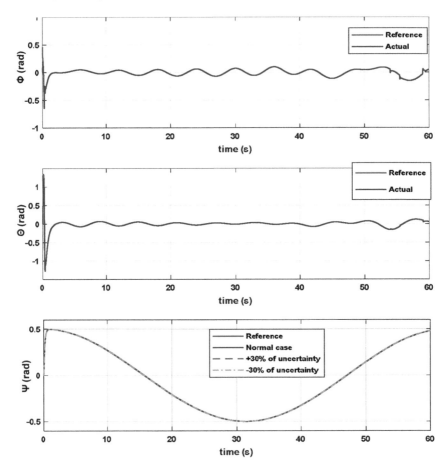

Figure 4.5   Attitude (Φ,Θ,Ψ)

and external perturbations. To validate the efficacy of the suggested solution, both theoretical analyses based on Lyapunov theory and Processor-in-the-loop experiments using an STM32F429 board have been thoroughly presented. Improved tracking precision, quick stabilization, and good tolerance to the unknown uncertainties, have all been guaranteed by the proposed approach. In order to eliminate the requirement of full-state measurements, a sliding mode observer will be designed in our future research. Real-time implementation of the developed controllers on a real quadrotor will be also investigated.

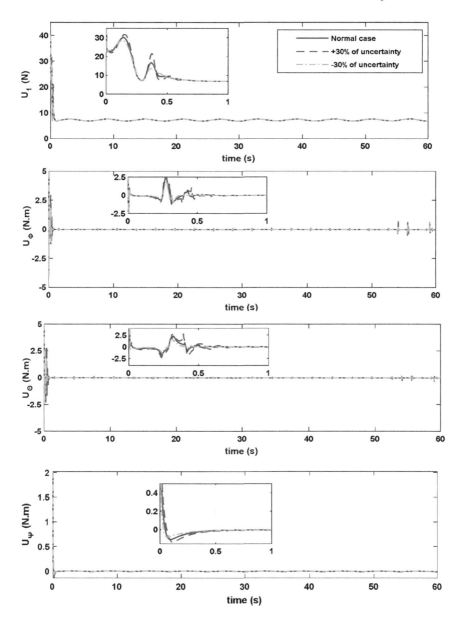

Figure 4.6 Control signals.

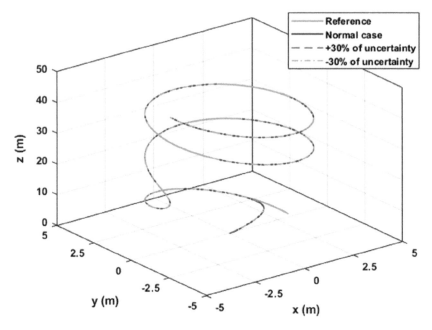

Figure 4.7   3D-path following under disturbances/uncertainties.

References

[1]   M. Hassanalian and A. Abdelkefi, "Classifications, applications, and design challenges of drones: A review," *Progress in Aerospace Sciences*, vol. 91, pp. 99–131, 2017.

[2]   H. Shraim, A. Awada, and R. Youness, "A survey on quadrotors: Configurations, modeling and identification, control, collision avoidance, fault diagnosis and tolerant control," *IEEE Aerospace and Electronic Systems Magazine*, vol. 33, no. 7, pp. 14–33, 2018.

[3]   H. Hassani, A. Mansouri, and A. Ahaitouf, "Robust autonomous flight for quadrotor UAV based on adaptive nonsingular fast terminal sliding mode control," *International Journal of Dynamics and Control*, vol. 9, no. 2, pp. 619–635, 2021.

[4]   H. Liu, J. Xi, and Y. Zhong, "Robust attitude stabilization for nonlinear quadrotor systems with uncertainties and delays," *IEEE Transactions on Industrial Electronics*, vol. 64, no. 7, pp. 5585–5594, 2017.

[5]   O. Mofid and S. Mobayen, "Adaptive sliding mode control for finite-time stability of quad-rotor UAVs with parametric uncertainties," *ISA transactions*, vol. 72, pp. 1–14, 2018.

[6] H. Hassani, A. Mansouri, and A. Ahaitouf, "A new robust adaptive sliding mode controller for quadrotor UAV flight," in *2020 IEEE 2nd International Conference on Electronics, Control, Optimization and Computer Science (ICECOCS)*, 2020, pp. 1–6. doi: 10.1109/ICECOCS50124.2020.9314413.

[7] K. Eliker, S. Grouni, M. Tadjine, and W. Zhang, "Quadcopter nonsingular finite-time adaptive robust saturated command-filtered control system under the presence of uncertainties and input saturation," *Nonlinear Dynamics*, vol. 104, no. 2, pp. 1363–1387, 2021.

[8] L. El Hajjami, E. M. Mellouli, and M. Berrada, "Robust adaptive non-singular fast terminal sliding-mode lateral control for an uncertain ego vehicle at the lane-change maneuver subjected to abrupt change," *International Journal of Dynamics and Control*, pp. 1–18, 2021.

[9] O. Elhaki and K. Shojaei, "A novel model-free robust saturated reinforcement learning-based controller for quadrotors guaranteeing prescribed transient and steady state performance," *Aerospace Science and Technology*, vol. 119, p. 107128, 2021.

[10] S. Ullah, Q. Khan, A. Mehmood, S. A. M. Kirmani, and O. Mechali, "Neuro-adaptive fast integral terminal sliding mode control design with variable gain robust exact differentiator for under-actuated quadcopter UAV," *ISA transactions*, 2021.

[11] N. P. Nguyen, N. X. Mung, H. L. N. N. Thanh, T. T. Huynh, N. T. Lam, and S. K. Hong, "Adaptive sliding mode control for attitude and altitude system of a quadcopter UAV via a neural network," *IEEE Access*, vol. 9, pp. 40076–40085, 2021.

[12] M. Navabi, A. Davoodi, and H. Mirzaei, "Trajectory tracking of an under-actuated quadcopter using Lyapunov-based optimum adaptive controller," *Proceedings of the Institution of Mechanical Engineers, Part G: Journal of Aerospace Engineering*, p. 09544100211010852, 2022.

[13] U. Tilki and A. C. Erüst, "Robust Adaptive Backstepping Global Fast Dynamic Terminal Sliding Mode Controller Design for Quadrotors," *Journal of Intelligent & Robotic Systems*, vol. 103, no. 2, pp. 1–12, 2021.

[14] H. Hassani, A. Mansouri, and A. Ahaitouf, "Adaptive Fast Terminal Sliding Mode Control for Uncertain Quadrotor Based on Butterfly Optimization Algorithm (BOA)," in *WITS 2020*, Springer, 2022, pp. 353–364.

[15] F. Muñoz, I. González-Hernández, S. Salazar, E. S. Espinoza, and R. Lozano, "Second order sliding mode controllers for altitude control of

a quadrotor UAS: Real-time implementation in outdoor environments," *Neurocomputing*, vol. 233, pp. 61–71, 2017.

[16] T. Jiang, T. Song, and D. Lin, "Integral sliding mode based control for quadrotors with disturbances: Simulations and experiments," *International Journal of Control, Automation and Systems*, vol. 17, no. 8, pp. 1987–1998, 2019.

[17] M. Labbadi and M. Cherkaoui, "Robust adaptive nonsingular fast terminal sliding-mode tracking control for an uncertain quadrotor UAV subjected to disturbances," *ISA transactions*, vol. 99, pp. 290–304, 2020.

[18] O. Mofid, S. Mobayen, C. Zhang, and B. Esakki, "Desired tracking of delayed quadrotor UAV under model uncertainty and wind disturbance using adaptive super-twisting terminal sliding mode control," *ISA transactions*, 2021.

[19] L.-X. Xu, H.-J. Ma, D. Guo, A.-H. Xie, and D.-L. Song, "Backstepping sliding-mode and cascade active disturbance rejection control for a quadrotor UAV," *IEEE/ASME Transactions on Mechatronics*, vol. 25, no. 6, pp. 2743–2753, 2020.

[20] A. Castillo, R. Sanz, P. Garcia, W. Qiu, H. Wang, and C. Xu, "Disturbance observer-based quadrotor attitude tracking control for aggressive maneuvers," *Control Engineering Practice*, vol. 82, pp. 14–23, 2019.

[21] L. Besnard, Y. B. Shtessel, and B. Landrum, "Quadrotor vehicle control via sliding mode controller driven by sliding mode disturbance observer," *Journal of the Franklin Institute*, vol. 349, no. 2, pp. 658–684, 2012.

[22] H. Hassani, A. Mansouri, and A. Ahaitouf, "Control system of a quadrotor UAV with an optimized backstepping controller," in *2019 international conference on intelligent systems and advanced computing sciences (ISACS)*, 2019, pp. 1–7.

[23] H. Hassani, A. Mansouri, and A. Ahaitouf, "Mechanical Modeling, Control and Simulation of a Quadrotor UAV," in *International Conference on Electronic Engineering and Renewable Energy*, 2020, pp. 441–449.

[24] H. Hassani, A. Mansouri, and A. Ahaitouf, "Modeling and Trajectory Tracking of an Unmanned Quadrotor Using Optimal PID Controller," in *WITS 2020*, Springer, 2022, pp. 457–467.

[25] B. Zhao, B. Xian, Y. Zhang, and X. Zhang, "Nonlinear robust adaptive tracking control of a quadrotor UAV via immersion and invariance

methodology," *IEEE Transactions on Industrial Electronics*, vol. 62, no. 5, pp. 2891–2902, 2014.

[26] H. Hassani, A. Mansouri, and A. Ahaitouf, "Robust hybrid controller for quadrotor UAV under disturbances," *International Journal of Modelling, Identification, and Control*.

[27] S. Motahhir, A. El Ghzizal, S. Sebti, and A. Derouich, "MIL and SIL and PIL tests for MPPT algorithm," *Cogent Engineering*, vol. 4, no. 1, p. 1378475, 2017.

[28] Y. Mazzi, H. Ben Sassi, F. Errahimi, and N. Es-Sbai, "PIL Implementation of Adaptive Gain Sliding Mode Observer and ANN for SOC Estimation," in *International Conference on Artificial Intelligence & Industrial Applications*, 2020, pp. 270–278.

[29] O. Mechali, J. Iqbal, X. Xie, L. Xu, and A. Senouci, "Robust Finite-Time Trajectory Tracking Control of Quadrotor Aircraft via Terminal Sliding Mode-Based Active Antidisturbance Approach: A PIL Experiment," *International Journal of Aerospace Engineering*, vol. 2021, 2021.

[30] N. Ullah *et al.*, "Processor in the Loop Verification of Fault Tolerant Control for a Three Phase Inverter in Grid Connected PV System," *Energy Sources, Part A: Recovery, Utilization, and Environmental Effects*, pp. 1–17, 2021.

[31] "Embedded Coder Support Package for STMicroelectronics Discovery Boards Documentation." https://www.mathworks.com/help/support-pkg/stmicroelectronicsstm32f4discovery/index.html (accessed Feb. 16, 2022).

SECTION 3
Smart Embedded Systems in Biomedicine

5

A Detailed Review on Embedded Based Heartbeat Monitoring Systems

S. Darwin[1], E. Fantin Irudaya Raj[*2]

[1]Assistant Professor, Department of Electronics and Communication Engineering, Dr. Sivanthi Aditanar College of Engineering, Tamilnadu, India
[*2]Assistant Professor, Department of Electrical and Electronics Engineering, Dr. Sivanthi Aditanar College of Engineering, Tamilnadu, India
[1]darwin.me111@gmail.com; [2]fantinraj@gmail.com

Abstract

In medical applications, embedded system technologies have become increasingly significant. Developing tools to improve the safety of healthcare professionals in the occurrence of epidemic contagious diseases, such as pandemic influenza, is a top priority. There has been an increase in demand for telemedicine services in recent years due to the rise in infectious diseases such as the Covid-19 viral infection. Telemedicine services include diagnostic tests, prognosis, and patient monitoring. Various heartbeat monitoring systems have been introduced to mitigate human distractions and explore the things that happen inside the human body. The cardiovascular system pushes blood and transports oxygen and nutrients via blood circulation. Heartbeat monitoring plays a vital role in preventing real-time accidents and providing the status of the heart's pumping prior to the external world. This review starts by using various sensors that measure the heart rate in a sophisticated manner, including invasive and non-invasive technologies. The amendable support with this modern device is retrofitted with the environmental conditions to provide the information rate without any delay. Nowadays, the emerging wireless sensor devices simplify collecting data regarding heart rate monitoring and sharing the information with anyone who resides in any part of the world. The types of sensors and the practical difficulties that various monitoring system faces are reviewed in detail.

5.1 Introduction

Embedded system applications are growing in popularity in recent times. The area of application varies from the device designs, garments, industries, healthcare, and armored vehicles, and handheld devices, but also in application areas such as mobile networks and 'e-worlds,' Artificial Intelligence, and IoT (Internet of Things), which allow for the creation of a wide range of software. Monitoring patients' heartbeat in remote houses is essential to take proper care after vacating the hospitals. The heartbeat activity in terms of electrical parameters is measured using Electrocardiography (ECG). Signal examinations using an electrocardiograph (ECG) can identify various traits in a patient's heart. Irregularities, chamber size and location, tissue injury, cardiac diseases prevalent, and heart rate are all characteristics. The issue with today's ECG signal equipment is that they can't characterize the data without the need for a full assessment and diagnosis by a specialist [1].

An ECG signal with various features is shown in Figure 5.1. A P wave is preceded by a QRS complex in a regular rhythm or cycle. The beat then comes to a halt with something like a T wave. A U wave can develop following a T wave on rare occasions.

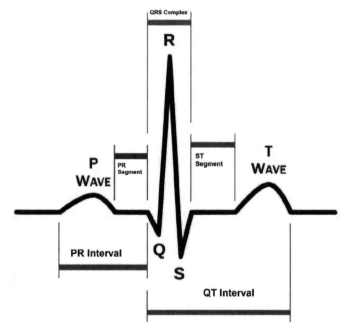

Figure 5.1   Signal waveform of ECG.

A physician in your area could do an ECG check. Typically, an ECG measurement tool has twelve leads. The cardiovascular system pumps blood around the body, transporting oxygen and other nutrients via systemic circulation. The human heart is seen in Figure 5.2. The upper left and right atria and the lower left and right ventricular are separated into four compartments.

The ECG is measured using electrodes placed on the body to obtain the heart's electrical signal, which has a few drawbacks. It is essential to monitor the person's heartbeat while working in a real-time environment. For example, while the person driving the car, workers in coal mines, oil plants, pilots, and astronauts have continuous monitoring of the beat to save guard human lives. As a result, traffic accidents due to heart problems appear to be a significant issue. In Mississippi, for instance, a school bus collapsed, injuring multiple kids when the coach driver encountered "sudden cardiac arrest" [2]. When a driver's physiological health deteriorates to the point where they lack control of the vehicle, technologies have been devised to execute an emergency evacuation for seizing control back to the driver [3]. In some environments, the emergency system based on embedded components is equipped. Modern, well-developed vehicles have guidelines concerning the healthcare of working people. It specifies it as part of routine management procedures, instruments, and software to monitor the employees' fitness are essential. For instance, the future advanced technologies not only monitor the activities of the workers but also take control over the machinery to prevent unwanted accidents. Nowadays, the driver's availability and health status are also monitored in modern vehicles. Still, along with it, the implementation of proper

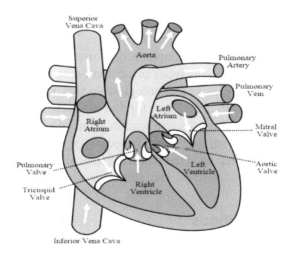

Figure 5.2 Signal waveform of ECG.

control over the vehicle without the vehicle driver's involvement in case any issues arise in the person's health.

Various monitoring systems are available, like biometric-based data collection of internal human health activities, including blood pressure and fat level with multiple parameters such as the pattern of sleep, diet, and exercise levels. The datum must be evaluated and analyzed more appropriately. The momentary heartbeat, which may be computed from the ECG, is a crucial piece of data. The transient heart rate fluctuates within a particular amount of variation in the resting condition. The variability in terms of heart rate is measured in a sophisticated way. The variable heart rate analysis allows for assessing cardiac autonomic stress, diagnosing angina and ischemic cardiovascular attacks, and classifying various illnesses [4].

The heart rate variation reflects the equilibrium between sympathetic nerves [5]. The sympathetic stimulation and reduced nerve (vagal) movement are considered. The spectrum analysis uses low-frequency component values of around 0.4Hz to 0.15Hz to depict the waveform indicating the person's attention. Usually, Fast Fourier Transform (FFT) is calculated using two R-R intervals, and the power spectral density value evaluates the low/high-frequency components. Hence complex signal processing of algorithms needs to be assessed the J peak signals of ballistocardiogram depending on the respiratory exertion and spurious movements that impede heartbeat [6]. Cardiopulmonary monitoring utilizes electrodes attached to the skin, which is uncomfortable. To avoid this, non invasive methodologies like Doppler radars are adopted [7]. The monitoring of respiratory action while traveling in vehicles is complicated and affects the frequency values; the efficiency of recognizing the pulse duration erodes the power ratio value of the signal-noise ratio to the minimum value [8].

The work presented in this chapter is organized in the following manner. Section 2 discusses the various types of equipment to monitor the heartbeat using invasive and non-invasive technologies. Section 3 provides an overview of the heartbeat detection algorithms for fast and efficient object measurements. Section 4 focuses on the application point of view of the heartbeat detection system and its software side. In section 5, the discussion about the hardware components that need to be utilized in heartbeat detection monitoring. The future work and conclusion of this review are discussed in section 6.

5.2 Measuring Methods

There are various traditional methods to monitor the heart rate in contact-oriented with the human body, like wearable equipment such as ring-type, available with cloths. As a result, even though the validity of the collected data is inferior to that of wearable electronics, these approaches are suitable for measuring the heart in a car, and a non wearable-type surveillance system is preferred. Without making contact with the person, the required details are collected.

The electric signals created by the body during every ventricular contraction are measured in the method used by conventional ECG. The magnitude of the heart's pumping behavior over time is used to evaluate the ECG. In twelve-lead ECG equipment, ten electrodes are employed. A total of six electrodes are positioned across the chest. The left/right arm and left/right leg are the four remaining electrodes on the limbs. The drawback of this conventional method is when electrodes get slack while the skin is moist, they are difficult to place correctly and are impacted by physical movement. The cardiogram-based measurement depends on the sound produced by the heart's beat. The finger-based equipment such as a stethoscope or microphone is used to evaluate the heart rate. The weak side of this measurement is the readings are disturbed by finger movements.

The sphygmomanometer is a technique for determining variations in arterial blood pressure as a function of heart pulse. The weak point is that the displacement of the human body affects the value. A photodiode or similar device receives the reflected light from near-infrared light delivered to the human skin to evaluate the value of heart pulse through wearing the equipment like a smartwatch. The readings are affected by physical movements, tactile state of fingers, skins, and other factors.

These approaches are acceptable for measuring the person's heartbeat in a vehicle or aircraft, and a non-wearable-type surveillance system is preferable to wearable computing. However, the gathered data is less reliable. Without establishing touch with the particular person, such tracking devices must accurately determine the driver's state of awareness. Table 5.1 depicts

Table 5.1 Invasive and non-invasive systems.

Technology	Evaluating methods	Applications	Targeted Source
Invasive	Watch type Belt type necklace type Ring-type	Emotion, stress [12,13], Drowsiness [14], Detecting respiratory symptoms [15, 16], heart beat monitoring [17,18], Respiratory Sinus arrhythmia [19]	Vehicle Drivers, Pilots/ astronauts, Patients with heart problems
Non-invasive	Camera type, Radio Frequency signals		

the various invasive and non invasive technologies. ISO 80601-2-69:2014 establishes standards for an oxygen concentrator's fundamental protection and functionality and its peripherals used to raise the oxygen level of gas meant for an individual patient. ISO 5367:2014 establishes the minimum standards for respiratory devices and tubes with anesthetic respiration systems and respiratory ventilation systems like nebulizers. IEC 62304 is an applicable regulatory standard that addresses software design and management. It outlines the procedures, actions, and tasks that must be followed to ensure safety. It is helpful for the establishment and improvement of medical device software. Hence, most of the devices met the regulatory standards with efficient patient measuring values to enhance the healthcare monitoring system.

The present chapter discussed the prerequisites for heart-rate embedded-based sensing utilized for various medical applications and heart rate monitoring systems. In addition to that, information about non invasive and invasive-based technologies was also revealed. Furthermore, the potential for building heart-rate sensing devices is discussed, considering the current ubiquitous use of handsets and smartwatches.

5.3 Categorisation of Algorithms for Heartbeat Detection

The spectrogram of the acquired Doppler radar information as a linear combination of source signals includes heartbeat, breathing, and physical movement, isolating only the heartbeat element using non-negative feature extraction and then sparse vector rejuvenation to make precise pulse rate predictions [9]. The evaluation, meanwhile, wasn't conducted for a pilot in a speeding object. Non-contact sensor nodes not only provide greater mobility and eliminate the need for connecting or cleanup electrodes, but they also have the special ability to be used on individuals suffering from skin problems, severe skin damage such as lacerations or burns, as well as patients who are anxious or allergic to contact sensors [10, 11]. Even when body movement noise was present, the heartbeat could be identified, allowing for assessing a constant pulse for a motorist in a speeding vehicle. The enhanced signal detection with the amalgamation of frequency analysis and differential method. By integrating multi-resolution investigation with quick data signal analysis and variations of the Hilbert transform as well as the Hamilton and Tompkins algorithms, the research was able to reach exceptional quickness and high accuracy for detecting R-wave peaks. Using a multi-resolution assessment algorithm based on Hilbert transform and main differential, the R-wave top can be found after removing characteristics based on the other parameters such as age, sex, level of heartbeat activity, and some other noise factors.

The previous methods relied on simple filtering of heartbeat-related data and imposing a threshold to the filtered signals to recover heartbeat locations [23] or spectral analysis from estimating heart rate frequency [24]. These methods could not provide both quick and actual performance and good prediction. Simple bandpass filtering would give a fast response time, but the filtered output signals would require additional computation to retrieve heart rhythm autonomously. As a result, the resilience of these procedures is severely hampered. To improve the accuracy nowadays for the retrieval of typical heart characteristics, the ensemble empirical mode decomposition (EEMD) [25] and the auto-correlation and frequency-time phase regression (FTPR) algorithms developed a noise-resistant technique [26]. The researchers in [27-28] have studied the fast Fourier transform (FFT), or the transform based on wavelet aspects (WT) were utilized to analyze a temporal fluctuation of the timeframe. The poly-phase basis discrete cosine transform was applied for pulse rate assessment [29]. The short-time Fourier transform (STFT) analysis in [30] was used to extract a specific heartbeat waveform, which was then filtered by an adaptive bandpass filter for increased precision. The insights obtained from the time - the domain of the cardiac signal on windows of 2–3 s were used to control the adaptive bandpass filter. Machine learning-based algorithms were utilized for computations [31]

The radar approach has evolved to be one of the most effective alternatives for non-invasive vital sign surveillance. It can produce small, minimal sensors that are fully non-obstructive and safe for human safety. Heart rate is extracted from discrete-time radar signals using different signal processing algorithms. Radar sensors are performed to recognize sub millimeter changes of the chest wall skin surface caused by pulse rate. The radar system has shown considerable promise in estimating heart rate and obtaining ventricular ejection period utilizing nonlinear filtering approaches.

Many research organizations have looked into the use of CW Doppler radars to detect a heartbeat. The majority of earlier study was obtained from experimental information recorded from healthy volunteers lying or sitting in a confined space. Figure 5.3 depicts the quadrature and in-phase signals of Doppler radar. Figure 5.4 illustrates the recorded signals of ECG and radar signals for the time period of 50–250ms.

Table 5.2 plots the various methods to classify the signals with the reference ECG. The author [32] has mentioned the retrieval of information and organization with the help of the ANN concept for more than 21 people with much less window time of less than 1 sec was revealed from the chart, which

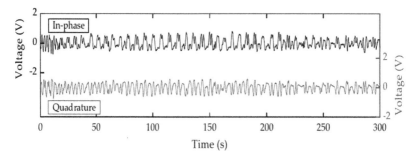

Figure 5.3 Quadrature and in-phases signals (Doppler radar).

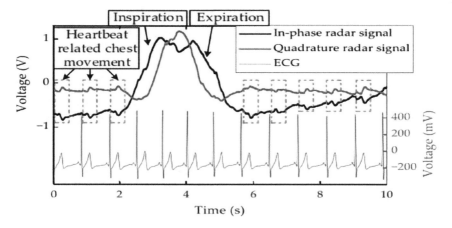

Figure 5.4 ECG and radar recorded signals for a single fragment.

enhances the performance of the device and produces a better result compared to other methods.

5.3.1 Categorisation of heartbeat displacement models

The radar signals had a characteristic signal form with a temporal latency in reference to the R wave of an ECG. Even during a heartbeat, the mechanical movement of the chest wall causes the heart rate disruption in the RF signals. While breathing and motion, this typical signal form became twisted and was diminished. Table 5.3 shows the various displacement waveform structures for processing the signals. An array of two successive pulses has a slight advantage compared to the other signals. Even though in the case of sine waves, few problems arise, such as distance detection and vibration in terms of amplitude.

Table 5.2 Methods for heartbeat retrieval.

Method	Radar Frequency (GHz)	Window Time (sec)	Total period (sec)	Average Error rate in percentage (%)
Classification with multiple signals [24]	2.4	8–27	30	~10
Decomposition ensemble empirical model [28]	5.8	15	240	3.67
Transform based on wavelet method [27]	5.8	3.5	60	3
Discrete Cosine Transform [22]	10.225	2	90	7.6
Regression method [21]	2.4	10–15	180	2
Chirp and Filter bank method [26]	24	3.5	300	1.54
Non-negative factorization matrix [30]	24	8	120	3.93
Filter Gamma[28]	5.8	15	600	3.8
Window variation based on time [33]	5.8	2–5	30	3.3
Artificial neural networks [32]	24	Less than 1	200	15

Table 5.3 Methods of heart monitoring displacement waveforms.

Reference	Waveform shape
[20]	An array of two succeeding pulses
[21]	Gaussian pulses
[22]	Half sine pulses
[34]	Sine wave

5.3.2 Demodulation techniques

Previously, a small-angle approximation method was used to demodulate the phase changes generated by the object's motion and is designed to check a "weak" vibration whose magnitude is much lower than the wavelength of the CW carrier [35]. Since the outcome of CW Doppler radar receivers is controlled by target motion via a sinusoid, it has two issues with detectable range and vibration amplitude. In the distance-dependent small-angle approximation method, the first is distance-dependent: that is, the radar/target distance varies, leading to deploy the receiver sensitivity, resulting in "optimum" and "null" points revealing the best/worst receiver performance in the aspect of the noise factor. Several strategies have been presented and studied to address this problem.

In the receiver, quadrature demodulation was implemented, allowing at least one channel which is not functioning at the null point to be selected for demodulation [36]. A frequency tuning was used, as well as a double-side-band transmission. The severe null point problem was alleviated by choosing the appropriate frequency spacing [37]. A voltage-controlled RF phase shifter was used to regulate the communication delays, which is comparable to adjusting the detecting range. However, further expenses of system complexity and adjustable feedback loop must be made to keep the sensor continuously functioning at the optimal position without incorporating distance dependencies [34].

When the comparison was made between the wavelength of the carrier and the amplitudes due to vibration, in this case, a small approximation was invalid. Even when the assessment is accomplished at the optimum location, powerful nonlinear harmonics and inter-modulation products will be formed, which do not represent the true mobility of the target. Furthermore, the amplitude characteristics of the vibration under test are not immediately extracted from the observable data in the small-angle approximation technique. One proven way is to use Bessel function expansion to estimate the baseband signal's harmonic ratio [38]. This strategy can only be used when a small number of vibration tones are available. There is also a high demand that the harmonic ratios be precisely determined. An arctangent demodulation technique has been developed to deal with the challenges [39]. The co-domain range of arctangent value is ($-\varpi/2$ to $\varpi/2$). A disconnection will happen once demodulation surpasses this range. In theory, phase unwrapping methods that relocate the discontinuity point by a multiple (integer) of ϖ can eradicate such discontinuity. However, deciding which point has to be shifted is almost impossible for hardware (or software), particularly when the amplitude of vibration is high. However, if the vibration amplitude is minimal, remote calibration can be challenging during the presence of noise. Due to the existence of noise in practice, this type of phase unwrapping may not always be beneficial [40]. Signal processing methods in optical communication utilize Differentiate and cross multiply (DACM), a much better modulation technique related to angle in optical communication. DACM may potentially eliminate the major limitations of the approximation method. As a result, it may be a viable choice for CW Doppler sensors.

Figure 5.5 depicts the measured I and Q values along with the detailed view of the arctangent function in Figure 5.5 (c), which is outdated; the discontinuity problem and co-domain issues are overcome with the assistance

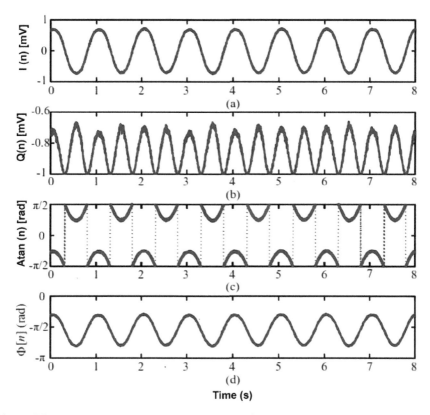

Figure 5.5 Measured values of optimum and null point values under a) I channel, b) Q channels, c) arctangent function, d) DACM function.

of a DACM that has been expanded. Irrespective of the detecting range, the frequency of [n] is always determined tobe1 Hz using the expanded DACM algorithm. In other words, in a small-angle approximation, it is distance independent of the null point condition.

5.4 Various Embedded Applications in the Medical Field

5.4.1 Software related to embedded systems

When dealing with embedded software, utilization of C programming language a lot. Whereas languages like java and C ++ are being used in particular sectors, and assembly is being used in hardware applications that require efficiency but lack a C compiler, no language apart from C is generally utilized. C seems to be the most popular language because it contains both middle-high functionalities, produces less code, is widely used in embedded

Table 5.4   Various embedded applications in the medical field.

Objective	Summary	Embedded systems
Fall detection system (invariant) [42]	Real-time invariant fall detection approach based on the Internet of Things. Individuals will be rescued, and medical aid will be provided through a notification system.	Accelerometer, Buzzer, GPS, GSM, Raspberry Pi, Node MCU, Arduino, Mobile phones
Gadgets utilisation to evaluate health status by measuring physiological parameters. [43]	Reduce error in false diagnosing of diseases Gadgets to examine and make decisions	Wearable devices like Apple watches, bio beat, biostrap
Monitoring pulmonary condition[44]	Mechanical ventilator with low cost, utilise respiratory frequency	Raspberry Pi, Arduino, servo motor, pressure sensor, monitor, resuscitator bag
Monitor respiratory cycle using IoMR [45]	Wearable device with the Internet of medical things	EMG sensors, Accelerometer, PPG as wearable one, Raspberry pi to analyse edge detection
Train the CNN deep learning algorithm to classify the X-rays of patients[46, 47]	Three layers of a framework (Hospital, Patient, Cloud)	x-ray dataset, cloud monitoring IoT
Temperature monitoring, mask identification, keeping social distance activities [48]	Detection algorithm for mask identification and social distance monitoring in home	Raspberry pi, Thermal camera, Arduino UNO, open CV, identifying mask-wearing algorithm, mobile-based application
Heartbeat detection using ANN[32]	The amalgamation of Doppler radar with ANN	Microcontroller with analog to digital converter, Radars
Face touch, sneeze, cough monitoring[49]	Activities like cough, sneezing, face touching can be monitored for the patients at home	Raspberry pi, Doppler radar, Fuzzy logic, application in mobile phones, Wi-Fi, CSTF monitor

computing libraries, is quite well, has a compiler for practically all micro-controllers, and has easy access to information. Table 4 plots out the various embedded application in the medical field. Older adults may have inadvertent injuries due to falls accidentally. The detection of falls can be monitored with support from the nodes. The nodes are interconnected to the servers and feed movements from the accelerometer to the servers in real-time. The server determines whether or not such a fall has transpired and reacts appropriately [42]. The gadgets utilization to monitor, diagnose the disease, and decide based on the algorithms [43]. The mechanical ventilators make use of raspberry pi boards to monitor the resuscitator bag [44], and wearable devices to examine the respiratory cycle of human beings [45]. Deep learning-based algorithms for classifying X-ray images [46] for efficient detection of diseases. In order to prevent airborne diseases from keeping social distance, monitoring has been examined with the help of embedded-based systems implemented [48]. The radar-based wireless healthcare monitoring systems have been implemented to remote monitor the patient from the healthcare center and the hospital [49–51].

Embedded system technology is improving at a quick pace as well. While previously dealing with microcontrollers with limited resources, it is now feasible to discuss devices that reach extremely high speeds. Subjects like internet protocols and encryption techniques have recently emerged as new research areas. We now require platforms that endorse us as programming becomes more complex. The Internet of Things (IoT) architecture includes "entry points," which interconnect other functionalities and objects to the network.

5.5 Embedded System Hardware for Heartbeat Monitoring

The collection of information from the above sections 3.1–3.4 is processed with the help of embedded hardware components such as microcontrollers, ASIC (Application Specific Integrated Circuits, FPGA (Field programmable gate arrays), and communication units. Microcontrollers, one of the most fundamental elements of embedded systems, are mono-chip computers. A microprocessor, memory, digital inputs and outputs, and other components are included.

The author [51] proposed the heartbeat monitoring system using the Atmega microcontroller-based embedded system, which operates at 8 MHz. Sensors, which are yet another hardware device in embedded devices, are developed to fulfill a variety of purposes as technology advances. The working subsystem consists mainly of an IR sensor used to sense and record the number of pulses over a particular duration by looking at the difference in the concentration of the blood circulation and evaluating the heart rate per minute. This information is fed to the microcontroller, which digitizes the acquired analog signal and displays the digital count on the LCD as a final output on its monitor.

The sensor's output signal is transferred to the microcontroller, which measures the pulse by transforming the analog signal to a digitized form and generating a control signal to produce the result in the Liquid Crystal Display (LCD). A timer is used to compute the heart rates, and the microcontroller's digitized output is relayed to the LCD. Figure 5.6 shows the working arrangement of the embedded-based heartbeat detection module.

A pulse sensor estimates the person's heartbeat when the person utilizes the device by placing the finger. The hardware used to compute heartbeat and the software used to monitor and maintain heartbeat data gathered in the previous phase make the embedded system [50]. This sensor is then connected to an Arduino Microcontroller, which allows the pulse rate to be checked and communicated to the network via the Bolt Wi-Fi device. The information is delivered via bolt cloud to an AWS server, which periodically analyses the pulse for any anomalies. When the patient's pulse surpasses the threshold, the client can establish a limit and then use online apps like Twilio to send text information to the doctor/guardian reporting the patient's clinical heartbeat, which is shown in Figure 5.7.

In [52], the same pulse sensor concept is used to evaluate the patient's heart rate. The microcontroller processes the collected data before sending it to the Wireless module (esp8266) for posting to the web servers. The data is analyzed and saved on a real-time basis, along with a time and date stamp from when it was recorded, which is illustrated in Figure 5.8.

Figure 5.9 depicts the non-invasive optical device that performs two basic functions: heartbeat detection and periphery capillaries oxygen saturation (Spo2). The projected and reflection modes are two separate methods

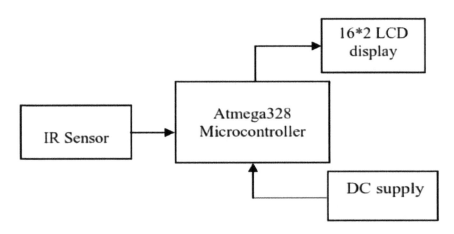

Figure 5.6   Embedded system with microcontroller.

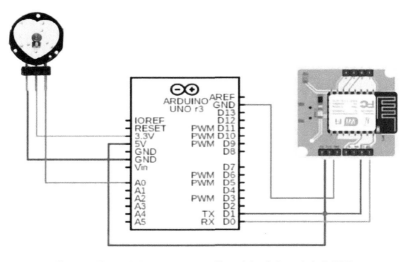

Figure 5.7 Arduino microcontroller with wi-fi module BOLT.

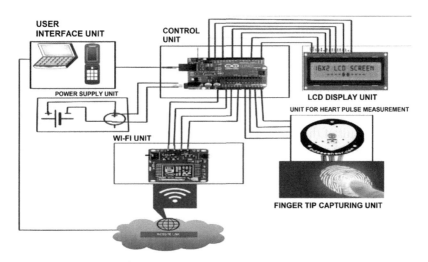

Figure 5.8 Arduino microcontroller with ESP8266 wi-fi device.

for estimating pulse rate and Spo2 readings. Then [53] has sensor works in reflection mode.

In this paper, [54-57] max30102 sensor is used to predict the pulse rate values. It consists of two lights; one is red, and the other is infrared, with wavelengths of 650nm and 950nm. Hence the absorption of light varies for oxygenated/deoxygenated blood. The sensor received the reflected backlight utilizing the algorithm the value of Spo2 was evaluated. To get blood from

Figure 5.9   Node MCU and arduino board connection to monitor heart rate / Spo2.

the heart to the rest of the body, it needs to be pounded at high pressure. It occurs as a result of the pulse, and it enables the arteries to be strained (blood pressure) when the blood flows through them. Arteries are the vessels that convey blood away from the heart. As a result of the blood pressure causing the artery to expand and shrink, the volume of the artery in the body parts (in this case, the fingertip) rises and falls. As the volume increases, more hemoglobin accumulates in the section region, raising the amount of consumed infrared light and decreasing the reflected signal back to the pulse rate sensor.

In Figure 5.10 Arduino Lilypad controller is used to estimate the heart rate wirelessly by means of sending SOS/to make calls using the GSM module of SIM900A [53]. A noise-reducing circuit sensor and a photonic integrated boosting circuit capture the information from the finger and deliver it to the microcontroller, which uses the logic to determine the heart rate. The RF module is used to transmit the messages or make calls for the guardian to intimate the situation of the person's current pulse rate [54].

In Figure 5.11, the non-invasive method of monitoring the person's respiratory system [55] is to evaluate the health without disturbing by means of wearing watches or any other equipment to collect the information. In section 3, the classification of heartbeat signals and demodulation techniques

Figure 5.10 Arduino lilypad with receiver section.

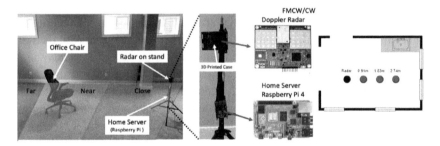

Figure 5.11 Respiratory system monitoring using raspberry pi as a home server.

were revealed. The utilization of Raspberry PI 4 as a home server and linked the radar to it through USB. The radar data is collected and processed by the home server. According to the ANN algorithm, the activities can be classified based on the respiratory values received from the phase-modulated signal.

5.6 Conclusion

The present state of the heart rate monitoring system was examined in the aspect of invasive and non-invasive technologies. The comprehensive study of the merits and demerits of the technologies revealed that the non-invasive mode plays a vital role in the future modern technologies without interfering or disturbing the person. The amalgamation of the non invasive method with cloud-based concepts suffices for the more accurate and sensitive device for heartbeat detection. The invasive technology-based methods' reliability mainly depends on the subject wearing the object; otherwise, there is no use. They are solely responsible for heart-rate detection and can be worn without constraining the driver. Furthermore, with the advent of smartphones in recent decades, cameras have become handheld, which is also a significant advancement. There are a number of issues with today's wristwatch and mobile phone technologies, as well as impediments to their practical implementation in vehicles. So non-invasive methods are widely used to assuage the drawbacks of the invasive type of technologies.

In this review, one of the upcoming robust radar-based non invasive technology was explored. The insinuate components such as the radar types, classification of algorithms from wavelet to Artificial Neural Networks, heartbeat displacement models, and demodulation techniques were examined. From the application point of view, the invasive and non invasive mode of technologies in the automobile sector to measure the heartbeat detection of the vehicle drivers to avoid accidents and revealed the merits and demerits of the technologies, which show the non-invasive method is one of the best contenders in the automobile field.

It is vital to examine the right hardware and applications by evaluating present and future technical developments in automated vehicles and also the recent developments in heartbeat detection and monitoring systems. By analyzing the situation based on the heartbeat of the vehicle's driver, the mechanical part alone takes control over the vehicle. Such technologies need a non invasive method of extracting the driver's heartbeat in much sophisticated and meticulously made computation with the support of Adaptive Neuro-fuzzy Inference System (ANFIS) type of algorithms for speedy calculations. In the future, the wireless-based sensor network will play a major role in healthcare due to its remote patient monitoring system. In the upcoming years, this non invasive technology, with the help of Radiofrequency wireless technologies, has greater potential to save and prolong the life span of many people. In this pandemic situation, the diseases are mostly spread through air borne, which can be mitigated with this kind of wireless sensor system technologies without any contact with the infected person and monitor the person remotely from the health care center itself.

References

[1]   Y. Hu, S. P. (1997, September). 'A patient-adaptable ECG beat classifier using a mixture of expert's approach,' IEEE Trans. Biomed. Eng., vol. 44, no. 9, pp. 891–900.

[2] CNN, Driver Killed and Seven Children Hurt in Mississippi School Bus Crash. Available online:https://edition.cnn.com/2019/09/10/us/mississippi-school-bus-crash/index.html (accessed on 30 July 2021).

[3] Kwon, S.; Jung, C.; Choi, T.; Oh, Y.; You, B. Autonomous Emergency Stop System. IEEE Intell. Veh. Symposium Proc. 2014, 444–449.

[4] Ryan, S. S., 'Understanding the Electrocadriogram (EKG or ECG) Signal', Retrieved December 2, 2014, from Atrial Fibrillation Resources for Patients: http://a-fib.com/treatmentsfor-atrial-fibrillation/diagnostic-tests/the-ekg-signal/

[5] J. H. Shin, S. H. Hwang, M. H. Chang, K. S. Park, Heart Rate Variability Analysis Using a Ballistocardiogram During Valsalva Manoeuvre and Post Exercise. Physiol. Meas. 2011, 32, pp.1239–1264.

[6] Y. Liu, Y. Lyu, Z. He, Y. Yang, J. Li, Z. Pang, Q. Zhong, X. Liu, H. Zhang, 'ResNet-BiLSTM: A Multiscale Deep Learning Model for Heartbeat Detection Using Ballistocardiogram Signals', J Healthc Eng., 2022,pp.6388445.

[7] P. Kontou, S. Ben Smida, S. Nektarios Daskalakis, S. Nikolaou, M. Dragone and D. E. Anagnostou, 'Heartbeat and Respiration Detection Using a Low Complexity CW Radar System,' 2020 50th European Microwave Conference (EuMC), 2021, pp. 929–932.

[8] K. Tsuchiya, K. Mochizuki, T. Ohtsuki, K. Yamamoto, 'Heartbeat Detection Technology for Monitoring Driver's Physical Condition,' SAE Technical Paper 2020-01-1212, 2020.

[9] C. Ye, K. Toyoda, T. Ohtsuki, 'Robust Sparse Adaptive Algorithm for Non-Contact Heartbeat Detection with Doppler Radar', IEICE Tech. Rep., 117, pp. 5–10, 2018.

[10] C. Gu, 'Short-range non-contact sensors for healthcare and other emerging applications: A review' Sensors 2016, 16, 1169.

[11] S. Izumi, Development of Non-Contact Heart Rate Variability and Respiration Monitoring Technology Using Microwave Doppler Sensor for in-Vehicle Application, Research Paper Funded by Takata Foundation; Takata Foundation: Tokyo, Japan, 2008.

[12] M. Zhao, F. Adib, D. Katabi, 'Emotion recognition using wireless signals' In Proceedings of the 22nd Annual International Conference on Mobile Computing and Networking, New York, NY, USA, 3–7 October 2016.

[13] J. A. Healey, R. W. Picard, 'Detecting stress during real world driving tasks using physiological sensors', IEEE Trans. Intell. Transp. Syst. 2005, 6, pp.156–166.

[14] B. G. Lee, B. L. Lee, W. Y. Chung, 'Wristband-type driver vigilance monitoring system using smartwatch', IEEE Sens. J. 2015, 15, pp. 5624–5633.

[15] L. Di Perna, G. Spina, S. Thackray-Nocera, M. Crooks, A. Morice, P. Soda, A. Den Brinker, 'An automated and unobtrusive system for cough detection', 2017, pp. 190–193.

[16] C. Hoyos Barcelo, J. Monge, Z. Pervez, L. San José Revuelta, P. Casaseca-de-la Higuera, 'Efficient computation of image moments for robust cough detection using smartphones', Computers in biology and medicine 100, 2018, pp.176–185.

[17] J. H. Choi, D. K. Kim, 'A remote compact sensor for the real-time monitoring of human heartbeat and respiration rate', IEEE Trans. Biomed. Circuits Syst. 2009, 3, pp.181–188.

[18] C. Li, Y. Xiao, J. Lin, 'Experiment and spectral analysis of a low-power Ka-band heartbeat detector measuring from four sides of a human body', IEEE Trans. Microw. Theory Technol. 2006, 54, pp.4464–4471.

[19] W. Massagram, V. Lubecke, A. Host-Madsen, O. Boric-Lubecke, 'Assessment of heart rate variability and respiratory sinus arrhythmia via Doppler radar', IEEE Trans. Microw. Theory Technol. 2009, 57, pp.2542–2549.

[20] V. L. Petrović, M. M. Janković, A. V. Lupšić, V. R. Mihajlović, J. S. Popović,-Božović,, 'High-Accuracy Real-Time Monitoring of Heart Rate Variability Using 24 GHz Continuous-Wave Doppler Radar', IEEE Access 2019, 7, pp.74721–74733.

[21] M. Nosrati, N. Tavassolian, 'High-accuracy heart rate variability monitoring using Doppler radar based on Gaussian pulse train modeling and FTPR algorithm', IEEE Trans. Microw. Theory Technol. 2018, 66, pp. 556–567.

[22] J. Park, J. W. Ham, S. Park, D. H. Kim, S. J. Park, H. Kang, S. O. Park, 'Polyphase-basis discrete cosine transform for real-time measurement of heart rate with CW Doppler radar', IEEE Trans. Microw. Theory Technol. 2018, 66, pp.1644–1659.

[23] B. K. Park, O. Boric-Lubecke, V. M. Lubecke, 'Arctangent demodulation with DC set compensation in quadrature Doppler radar receiver systems', IEEE Trans. Microw. Theory Technol. 2007, 55, pp.1073–1079.

[24] P. Bechet, R. Mitran, M. Munteanu, 'A non-contact method based on multiple signal classification algorithm to reduce the measurement time for accurately heart rate detection', Rev. Sci. Instrum. 2013, 84, 084707.

[25] W. Hu, Z.Zhao, Y. Wang, H. Zhang, F. Lin, 'Non-contact accurate measurement of cardiopulmonary activity using a compact quadrature Doppler radar sensor', IEEE Trans. Biomed. Eng. 2014, 61, pp.725–735.

[26] M. Nosrati, N. Tavassolian, 'High-accuracy heart rate variability monitoring using Doppler radar based on Gaussian pulse train modeling

and FTPR algorithm', IEEE Trans. Microw. Theory Technol. 2018, 66, pp.556–567.

[27] M. Li, J. Lin, 'Wavelet-transform-based data-length-variation technique for fast heart rate detection using 5.8-GHz CW Doppler radar', IEEE Trans. Microw. Theory Technol. 2018, 66, pp.568–576.

[28] J. Tu, J. Lin, 'Fast acquisition of heart rate in non-contact vital sign radar measurement using time-window-variation technique', IEEE Trans. Instrum. Meas. 2016, 65, pp.112–122.

[29] J. Park, J. W. Ham, S. Park, D. H. Kim, S. J. Park, H. Kang, S. O. Park, 'Polyphase-basis discrete cosine transform for real-time measurement of heart rate with CW Doppler radar', IEEE Trans. Microw. Theory Technol. 2018, 66, pp.1644–1659.

[30] K. Yamamoto, K. Toyoda, T. Ohtsuki, 'Spectrogram-based non-contact RRI estimation by accurate peak detection algorithm', IEEE Access 2018, 6, pp.60369–60379.

[31] J. J. Saluja, J. J. Casanova, J. Lin, 'A Supervised Machine Learning Algorithm for Heart-rate Detection Using Doppler Motion-Sensing Radar', IEEE J. Electromagn. RF Microw. Med. Biol. 2019, 4, pp.45–51.

[32] N. Malešević, V. Petrović, M. Belić, C. Antfolk, V. Mihajlović, M. Janković, 'Contactless Real-Time Heartbeat Detection via 24 GHz Continuous-Wave Doppler Radar Using Artificial Neural Networks', Sensors 2020, 20, 2351.

[33] C. Ye, K. Toyoda, T. Ohtsuki, 'Blind source separation on non-contact heartbeat detection by non-negative matrix factorisation algorithms', IEEE Trans. Biomed. Eng. 2019, 67, pp.482–494.

[34] J. Wang, X. Wang, L. Chen, J. Huangfu, C. Li, L. Ran, 'Non-contact distance and amplitude-independent vibration measurement based on an extended DACM algorithm', IEEE Trans. Instrum. Meas. 2014, 63, pp.145–153.

[35] W. Pan, J. Wang, J. Huangfu, C. Li, and L. Ran, Null point elimination using RF phase shifter in continuous-wave Doppler radar system', Electron. Lett., vol. 47, no. 21, 2011, pp. 1196–1198.

[36] A. D. Droitcour, O. Boric-Lubecke, V. M. Lubecke, J. Lin, and G. T. Kovacs, Range correlation and I/Q performance benefits in single chip silicon Doppler radars for non-contact cardiopulmonary monitoring,' IEEE Trans. Microw. Theory Tech., vol. 52, no. 3, 2004, pp. 838–848.

[37] Y. Xiao, J. Lin, O. Boric-Lubecke, and V. M. Lubecke, 'Frequency tuning technique for remote detection of heartbeat and respiration using low power double-sideband transmission in Ka-band,' IEEE Trans. Microw. Theory Tech., vol. 54, no. 5, 2006, pp. 2023–2032.

[38] Y. Yan, C. Li, J. A. Rice, J. Lin, 'Wavelength division sensing RF vibrometer," Proc. IEEE MTT-S, 2011, pp. 1–4.

[39] B. Park, O. Boric-Lubecke, and V. M. Lubecke, 'Arctangent demodulation with DC offset compensation in quadrature Doppler radar receiver systems,' IEEE Trans. Microw. Theory Tech., vol. 55, no. 5, 2007, pp. 1073–1079.

[40] K. Itoh, 'Analysis of the phase unwrapping problem,' Appl. Opt., vol. 21, no. 14, 1982, p. 2470.

[41] F. Schadt, F. Mohr, and M. Holzer, 'Application of Kalman filters as a tool for phase and frequency demodulation of IQ signals,' in Proc. IEEE Int. Conf. Comput. Technol. Electr. Electron. Eng., Jul. 2008, pp. 421–424.

[42] Nooruddin Sheikh, Islam Md, Sharna Falguni, 'An IoT based device-type invariant fall detection system', 2019. https://doi.org/10.1016/j.iot.2019.100130. 9.100130.

[43] D. R. Seshadri, E. V. Davies, E. R. Harlow, J. J. Hsu, S. C. Knighton, T. A. Walker, J. E. Voos, C. K. Drummond, 'Wearable sensors for COVID-19: a call to action to harness our digital infrastructure for remote patient monitoring and virtual assessments', Front. Digit. Health 2020;2(8).

[44] Acho Leonardo, N. Vargas Alessandro, Pujol Vazquez Gisela, 'Low cost, open-source Mechanical Ventilator with pulmonary monitoring for COVID-19 patients,' Actuator. MDPI 2020;9(3).

[45] Qureshi Fayez, Krishnan Sridhar, 'Wearable hardware design for the internet of medical things (IoMT),' Sensors 2018;18:3812.

[46] El-Rashidy Nora, El-Cappagh Shaker, S. M. Riazul Islam, M. Hazem, El-Bakry, Abdelrazek Samir, 'End to End deep learning framework for coronavirus. (COVID-19) detection and monitoring' 2020;9(1):pp.1–25

[47] Ch, Gangadhar, et al. "Diagnosis of COVID-19 using 3D CT scans and vaccination for COVID-19." *World Journal of Engineering* (2021).

[48] Petrovic Nenad, Kocić Đorđe, 'IoT-based system for COVID-19 indoor safety monitoring,' IcETRAN; 2020.

[49] Elishiah Miller, Nilanjan Banerjee, Ting Zhu, 'Smart homes that detect sneeze, cough, and face touching,' Smart Health, Vol. 19, 2021, 100170,

[50] R. Gatti, H. A. Roopashree, R. Pruthvika, K. Priyanka, S. Varun, 'Advanced Heart Rate Detection Using Embedded System,' 3rd IEEE International Conference on Recent Trends in Electronics, Information & Communication Technology (RTEICT), 2018, pp. 1488–1493,

[51] Neelakandan, S., et al. "Blockchain with deep learning-enabled secure healthcare data transmission and diagnostic model." *International*

Journal of Modeling, Simulation, and Scientific Computing 2022: 2241006.

[50] H. K. Pendurthi, S. S. Kanneganti, J. Godavarthi, S. Kavitha, H. S. Gokarakonda, 'Heart Pulse Monitoring and Notification System using Arduino,' 2021 International Conference on Artificial Intelligence and Smart Systems (ICAIS), 2021, pp. 1271–1278,

[51] S. Abba, A. M. Garba, 'An IoT-Based Smart Framework for a Human Heartbeat Rate Monitoring and Control System,' Proceedings 2020, 42, 36.

[52] Mustafa A Al-Sheikh and Ibrahim A Ameen, 'Design of Mobile Healthcare Monitoring System Using IoT Technology and Cloud Computing' 2020 IOP Conf. Ser.: Mater. Sci. Eng. 881 012113

[53] S. Mukherjee, A. Ghosh, S. K. Sarkar, 'Arduino based Wireless Heart-rate Monitoring system with Automatic SOS Message and/or Call facility using SIM900A GSM Module,' 2019 International Conference on Vision Towards Emerging Trends in Communication and Networking (ViTECoN), 2019, pp. 1-5.

[54] Deivakani, M., et al. "VLSI Implementation of Discrete Cosine Transform Approximation Recursive Algorithm." *Journal of Physics: Conference Series*. Vol. 1817. No. 1. IOP Publishing, 2021.

[54] E. Miller, N. Banerjee, T. Zhu, 'Smart Homes that Detect Sneeze, Cough, and Face Touching,' Smart Health, 2020, 100170 https://doi.org/10.1016/j.smhl.2020.100170.

[55] Stalin David, D., et al. "Inflammatory syndrome experiments related with COVID-19." *Turkish Journal of Physiotherapy and Rehabilitation* (2021): 765–768.

[56] Agarwal, Parul, et al. "Parameter Estimation of COVID-19 Second Wave BHRP Transmission Model by Using Principle Component Analysis." *Annals of the Romanian Society for Cell Biology* (2021): 446–457.

[57] Gnanasekar, A. K., et al. "Novel Low-Noise CMOS Bioamplifier for the Characterisation of Neurodegenerative Diseases." *GeNeDis 2020*. Springer, Cham, 2021. 221–226.

6

Embedded Systems in Biomedical Engineering: Case of ECG Signal Processing Using Multicores CPU and FPGA Architectures

Wissam Jenkal[1], Safa Mejhoudi[1], Amine Saddik[1], Rachid Latif[1]

[1]LISTI Laboratory, National School of Applied Sciences, Ibn Zohr University, Agadir, Morocco
w.jenkal@uiz.ac.ma

Abstract

This chapter proposes a survey on the use of embedded systems in some biomedical applications such as ECG signal processing. Various embedded architectures are proposed such as Multi-cores CPU, FPGA, and CPU-FPGA architectures. Some recently published implementations of ECG signal denoising methods are also proposed in this chapter. These implementations concern time-domain analyses as ADTF technique, time-frequency domain as DWT technique, or a hybrid approach as ADTF-DWT technique.

6.1 Introduction

Nowadays, embedded systems are emerging considerably in different engineering and research fields. This is due to their high possibility to offer specified needs to these fields. From very low-cost systems to high performances ones, embedded systems propose different architectures to respond to different criticalities of embedded approaches [1].

Embedded systems are complex systems that integrate software and hardware designed together to provide given functionality. They are generally composed of one or more microprocessors intended to execute various programs defined during conception and saved in memories. To optimize the performance and reliability of these systems, programmable digital circuits FPGA (Field Programmable Gate Array), ASIC circuits (Application Specific Integrated Circuits), or analog modules are additionally used [2].

Embedded systems generally operate in Real-Time (RT). Calculation operations are then performed in response to an external event (hardware interruption). The validity and relevance of a result depend on when it is delivered. A missed deadline induces an operating error that can cause either a system breakdown (crash) or a non dramatic degradation of its performance. In autonomous embedded systems, power consumption is a critical point for cost. Indeed, excessive consumption increases the cost price of the onboard system since high-capacity batteries are then required. This explains why the high optimization of embedded architectures can even reduce the deliverance time of the results of the system to respond to the real-time criticalities, as well as the total cost of a proposed system [3].

In current design strategies, an embedded system is generally integrated on single silicon support; thus, constituting a complete system integrated on an SoC chip or (System on a Chip). Systems on a chip usually contain a wide variety of programmable devices such as microcontrollers, DSP (Digital-Signal Processor) signals, and ASICs which are developed for complex applications [2].

Recently, embedded systems are widely proposed in different approaches to biomedical engineering. This is due to the positive and crucial impact brought by new technologies on human life [4]. The implementation and algorithmic optimization of different biomedical data analysis techniques in an embedded system can take an important role in the medical analysis or the monitoring of a patient's health status. This is based on the computing power and flexibility of embedded system architectures [1].

This chapter presents an overview of some proposed architectures based on different embedded systems technologies applied in biomedical applications, for example, ECG signal processing.

The ECG signal, shown in figure 6.1, represents the electrical activity of the human heart. This activity is recorded on a patient using electrodes placed on the top of the skin. ECG signal analysis is a major challenge for researchers and biomedical engineers. A great diversity of research and development axes are proposed in the state of the art. This varies between the analysis of the ECG signal (pre-processing of the ECG signal, extraction of the QRS complex, etc.), the implementation, and the algorithmic and hardware optimization of the techniques for analyzing this signal [5]. Figure 6.2 explains the working methodology chosen for all the work presented in this chapter.

The next section of this chapter presents an overview of some different embedded systems proposed in the state of the art. The third section presents

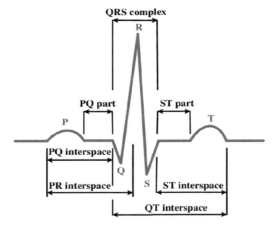

Figure 6.1   Example of an ECG signal with its features.

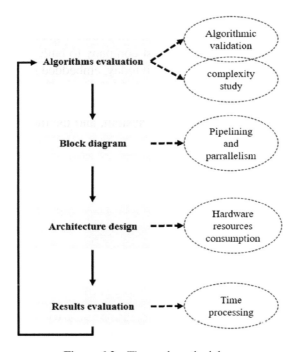

Figure 6.2   The work methodology

some embedded systems applications in biomedical engineering: the case of the pre-treatment of ECG Signal. Two axes are proposed in this section, the first one deals with some proposed techniques for embedded systems implementations. The second axe proposes the implementation of these approaches in different embedded systems technologies. Finally, the last section of this chapter presents a discussion and conclusions of the results mentioned in the previous part and the different perspectives accumulated from earlier research studies.

6.2 Embedded System Architectures

6.2.1 Embedded systems overview

Embedded systems have experienced a huge revolution in technology and the tools used for operating these systems. The most standard definition can be summarized in the fact that embedded systems are dedicated systems interconnected between software and hardware to have a system that exploits hardware resources for software problems. These systems can be in real-time by respecting the time constraints. These constraints differ from one specific application to another depending on the environmental requirements. Embedded systems were first used in 1960 for guidance applications. In fact, Texas Instruments were the first company to build a microcontroller for industrial applications in 1971. Until today, embedded systems have been developed into complex systems with multi-devices. Generally, embedded systems are divided into three categories: The first category is the homogeneous system, the second is the mixed system, and the third category is the heterogeneous system.

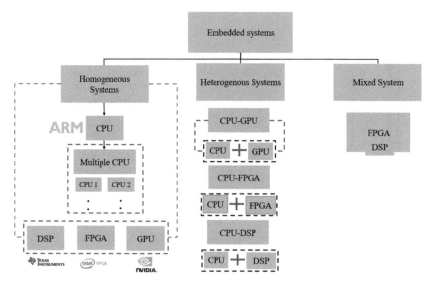

Figure 6.3 Different types of embedded systems.

Homogeneous systems are architectures based on processing units of the same type. We can identify four types: FPGAs, CPUs, GPUs, and DSPs. Moreover, the mixed architecture is generally based on constructing an embedded CPU system based on a hardware architecture like FPGAs. For heterogeneous systems, we can find CPU-GPU, CPU-DSP, and CPU-FPGA. These systems can also contain several devices by combining CPU-GPU/DSP or CPU/FPGA/DSP. Each system has its own language to exploit the whole resources. Regarding homogeneous systems, we can find C, C++, OpenMP, and VHDL for systems that require a hardware description. In mixed systems, we can find the use of system C. As for heterogeneous systems, we find OpenCL for all types of architecture: CPU-FPGA, CPU-GPU, or CPU-DSP. In addition to that, there is CUDA, which is a parallel programming language for heterogeneous architecture. But the problem with this language is that it can be used solely in Nvidia and CPU-GPU architecture. The following figure shows the different types of embedded systems.

6.2.2 Embedded systems architectures

6.2.2.1 Standalone architectures

CPU Architectures

A CPU or Central Processing Unit refers to a processor-one of the main elements that go into the composition of electronic devices. The processor has an executor role. It abides by the instructions given to it by computer programs. In fact, its function resembles that of the human brain.

The CPU comprises three main parts: control unit (CU), arithmetic logic unit (ALU), and registers [6]. The architecture of a CPU is presented in Figure 6.4.

The CU tries to find instructions in memory, decodes them, and then coordinates the rest of the processor to execute them. An elementary control unit consists of an instruction register and a "decoder/sequencer" unit. The ALU executes the logic and arithmetic instructions requested by the CU. The instructions relate to one or more operands. Execution speed is optimal when the operands are located in registers rather than in external memory to the processor. Registers are memory cells internal to the CPU. They are few but very quick to access. They are used to store variables, intermediate results of operations (arithmetic or logical), or processor control information.

The structure of the registers varies from one processor to another. This is what makes each type of CPU have its own instruction set. Their basic

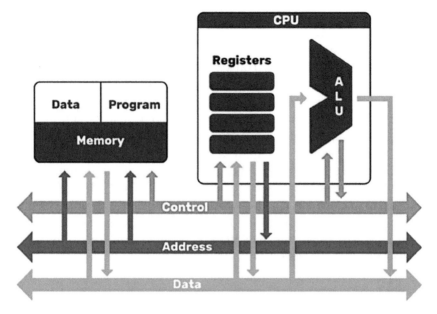

Figure 6.4   Architecture of a CPU.

functions are nevertheless similar, and all processors have roughly the same categories of registers:

- The accumulator is primarily intended to contain the data which the ALU must process.

- General registers are used to store intermediate results.

- Address registers are used to create specific data addresses. These are, for example, the base and index registers that make it possible to organize the data in memory as indexed tables.

- The instruction register contains the instruction code, which is processed by the decoder/sequencer.

- The program counter saves the address of the following instruction to be executed. In principle, this register keeps counting. It generates the addresses of the instructions to execute one after the other.

- The state register, sometimes called the condition register, contains indicators called flags whose values (0 or 1) vary according to the results of arithmetic and logic operations. These states are used by the conditional jump instructions.

- The stack pointer manages certain data in memory by organizing them in stacks.

The content of the program counter is placed on the addressing bus to search there for instructions in machine code. The control bus produces a read signal and the memory, which is selected by the address, returns the code of the instruction to the processor via the data bus.

Once the instruction lands in the instruction register, the control unit decodes it and produces the appropriate sequence of internal and external signals that coordinate its execution. An instruction consists of a series of elementary tasks, which are punctuated by clock cycles.

All the tasks that make up an instruction are executed one after the other. The execution of an instruction, therefore, takes several cycles. As it is not always possible to increase the frequency, the only way to increase the number of instructions processed in a given time is to seek to execute as many instructions as possible simultaneously [7]. This is accomplished by splitting up CPU resources, data, and/or processes. This is called parallelization.

Specialized CPU designs are now one of the most often used methods for achieving high performance, particularly in embedded systems. Some of the existing CPU implementations are currently incompatible with mixed-criticality systems with strict real-time requirements.

Real-time systems are also used in all the embedded applications in the social and economic sectors, including biomedical engineering [8]. We may argue that no area is complete without one or more microprocessors. As a result, research in this field has accelerated, resulting in significant advancements for challenging real-time applications. Among the proposed solutions is that of the multiprocessor architectures, which will be presented later in the section on heterogeneous architectures.

DSP Architectures

A DSP for "Digital Signal Processor" is a particular type of microprocessor. It is characterized by integrating a set of special functions that are intended to make it particularly efficient in digital signal processing.

A DSP is implemented by combining memory (RAM, ROM) and peripherals like a conventional microprocessor [9]. A typical DSP is more intended to be used in stand-alone processing systems. Therefore, it is generally in the form of a microcontroller incorporating, depending on the brands and ranges of manufacturers, memory, fast synchronous serial ports, timers, DMA controllers, and different I/O ports.

In general, the architecture of a digital signal processing system can be represented schematically in Figure 6.5, where the ADC is an analog-digital converter and the DAC is a digital-analog converter.

The block diagram of a typical DSP is presented in figure 6.6. The data and program memory blocks were firstly blocks outside the DSP, but as the level of integration increases, these blocks are now fit on-chip [10]. The data

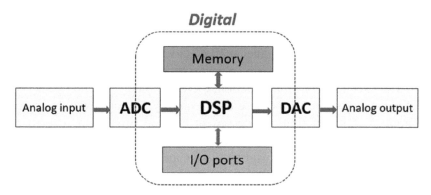

Figure 6.5   Typical chain of a digital signal processing system.

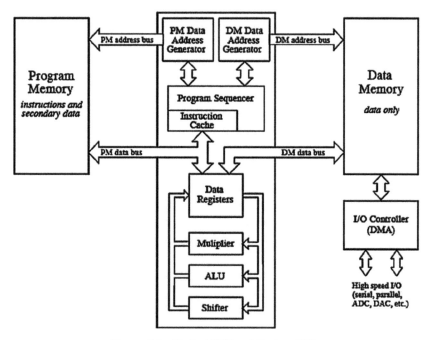

Figure 6.6   Typical DSP architecture [10].

is saved in this memory. One memory port is connected to the I/O controller which continuously accepts data from DACs and ADCs.

ADC takes the analog input and converts it into digital of proper width with a suitable sampling rate. The IO controller writes it into the data RAM. This data is saved in RAM after being processed by the DSP, where the IO controller will collect data and send it to the DAC. Analog output will be provided by the DAC. ADCs and DACs can also be incorporated into the DSP.

GPU Architectures

Graphic processors have grown significantly. Compared to CPUs, GPUs use many threads that copy between them to exploit the total parallelism of the architecture. This number of threads varies between architectures depending on the hardware's need and limitations.

GPUs have adopted SIMD (Single Instruction Multiple Data) as a basis for parallelism. The reason why SIMD is the best choice can be summarized in the fact that GPU threads are not as high in terms of processing frequency as CPU threads. For this reason, they are weak on the execution side of many mathematical instructions, which makes data parallelism more efficient. In general, the exploitation of GPU architecture requires specific languages like OpenGL to parallelize the different data or instructions in the GPU threads.

The Nvidia GPU architecture is extremely different from the CPU architecture. It consists of several core blocks:

- Multiprocessors, called "Streaming multiprocessors" (SM)

- Processors or "CUDA Core," called "Streaming processors" (SP)

- Memory (global, constant, shared)

A GPU consists of one or more multiprocessors. Each one of those multiprocessors has N CUDA cores (processors). The more multiprocessors you have, the more GPU will be able to process tasks simultaneously or the same task faster. The performance of Nvidia hardware has improved by combining an improvement in the number of multiprocessors and the number of cores per multiprocessor.

Each multiprocessor has access to what is called a register file, which is a block of memory that runs at the same speed as the processors (CUDA cores). Each processor has its own private register. In terms of memory, this results in zero latency which is a local memory for each processor.

Each multiprocessor has its own on-chip memory in the form of shared memory and an L1-level cache. All processors in one multiprocessor can access this memory while processors in another multiprocessor cannot.

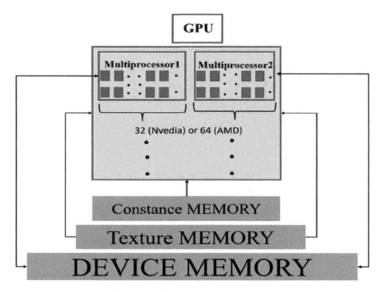

Figure 6.7 GPU architecture.

The L2 level cache is the data unification point between the different multiprocessors. It takes care of all load and store requests and provides faster access to the global memory allowing high-speed data sharing. All processors can access the texture memory and the constant memory used for read-only data. The GPU has a global memory. All processors in a multiprocessor have access to this memory. The following figure shows the architecture of a GPU.

FPGA Architectures

FPGAs (Field Programmable Gate Arrays or "programmable logic networks") are fully reconfigurable VLSI components, which allows them to be reprogrammed at will to significantly accelerate certain calculation phases.

The advantage of this type of circuit is its great flexibility which allows them to be reused at will in different algorithms in a very short time.

The progress of these technologies makes it possible to make even faster and more profound integrated components, which makes it possible to program important applications.

FPGA circuits consist of a matrix of programmable logic blocks surrounded by programmable input-output blocks. The assembly is linked by a programmable interconnection network.

FPGAs are quite distinct from other families of programmable circuits while offering the highest level of logic integration. Here is an example of the internal structure of asymmetric array type FPGA (figure 6.8):

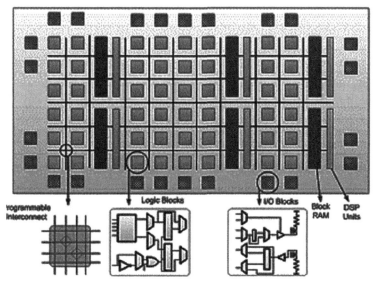

Figure 6.8   Example of FPGA internal structure.

Besides the outputs and inputs of the circuit, the advantage of FPGAs relies in their ability to be configured on-site without sending the circuit to the manufacturer, which allows them to be used for a few minutes after their design. The most recent FPGAs can be configured in a hundred milliseconds. FPGAs are used for the fast and inexpensive development of ASICs.

Hardware Description Language (HDL)

A hardware description language (HDL) is a specialized language for describing the architecture and behavior of electrical circuits, particularly digital logic circuits [11]. A hardware description language allows for the exact and formal design of an electronic circuit as well as automated analysis and simulations. It can also convert an HDL description into a Netlist, which may then be put and routed to form an integrated circuit.

The HDL consists of a textual description including operators, expressions, and input and output declarations. HDL compilers give a logic gate interconnection map instead of a computer-executable file. [12]. This resulting map is then loaded into the programming device to verify the operations of the desired circuit.

The HDL language aims to describe digital circuits in the form of structural, behavioral, and synthesis levels. The most commonly used HDL languages for synthesis today are VHDL and Verilog [13]. In this study, we propose the use of the VHDL language.

VHDL (VHSIC HDL) is a hardware description language used in electronic design automation to describe systems such as FPGAs. VHDL can also be used as a general-purpose parallel programming language.

Primary and secondary design units are included in the VHDL description. Entity and Package are the two main design units. Secondary design units, which make up the package's architecture and body, are always linked to a primary design unit. Primary and secondary design units are collected in libraries. One or more design unit libraries are frequently found in a typical design.

A VHDL entity specifies the entity name, entity ports, and entity information. All models are designed using one or more entities. The entity describes the interface of the VHDL model. It specifies the number, the direction, and the type of the ports.

The architecture specifies the entity's fundamental functionality and includes the instructions that model the entity's behavior. Architecture is always associated with an entity and specifies the behavior of that entity, which may be described by several architectures.

6.2.2.2 Heterogeneous architectures
Multi-CPU Architectures

Multicore means an architecture where a single hardware processor includes the core logic of several processors. A single chip is employed to wrap these processors. These single chips are known as processor chips. The multicore system architecture takes several processor cores and merges them into a single physical processor. The aim is to build a device that can handle more

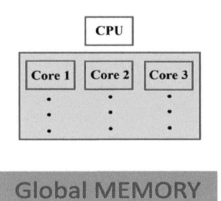

Figure 6.9   Multi-CPU architecture.

processing tasks simultaneously, leading to higher system efficiency. The following figure shows an overview of the multicore architecture.

CPU-GPU Architectures

CPU-GPU systems contain two parts, one for the host and the other for the device. The host part is always a CPU, but the device part is a GPU. The CPU part takes charge of initializing and sending the necessary data to the GPU part as well as sending commands to the GPU for the implementation of the different algorithms. This ordering is based on sending the number of threads to execute and the number of workgroups. Generally, the use of this system is done in two levels of parallelization. The first one is based on the interconnection of treatment between the CPU and the GPU to distribute the treatment tasks between them. This mode of parallelism is done using specific languages like CUDA, and OpenACC for Nvidia architectures. On the other side, we can find the use of OpenCL for different architectures. The second level of parallelism is based on the combination of OpenMP and OpenCL. This level of parallelism exploits the entire target architecture based on CPU cores using OpenMP and GPU threads using OpenCL. In this case, we can conclude that the use of these types of architecture is based on a thorough study to exploit the totality of the resources based on an optimal implementation. Additionally, these architectures have a memory model ecosystem. This memory model is based on the global memory of the host and device that communicate with each other through buses. The wrong communication between the CPU and GPU part can sometimes create memory latency problems which influence the execution time of such algorithms. The following figure shows the architecture of a CPU-GPU system.

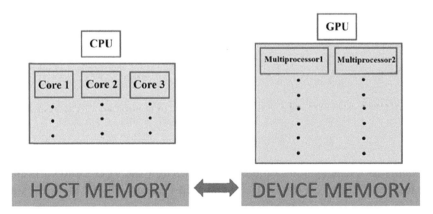

Figure 6.10 CPU-GPU architecture.

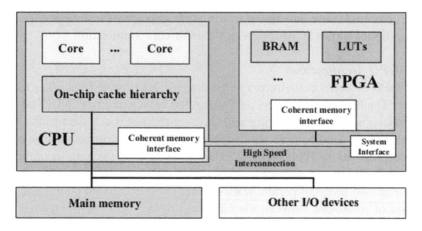

Figure 6.11   Target architecture of a heterogeneous system integrating CPU and FPGA.

CPU-FPGA Architectures

As accelerators continue to raise the bar for performance and energy efficiency, heterogeneous architectures combining CPU and FPGA are now becoming increasingly desirable for achieving significant performance gains. Because of developments in interconnection technologies between heterogeneous devices, data communication is also becoming much more efficient. Since the CPU and FPGA may interact through shared memory in these systems, the cache hit rate improves and overall data transmission latency decreases [14].

Figure 6.11 depicts a heterogeneous system platform that combines CPU and FPGA. To develop individualized hardware accelerators, the FPGA logic consists of look-up table logic and on-chip memory resources. The CPU is a multi-core general-purpose processor with a cache structure on the chip. For fast data transfer, the CPU and FPGA include coherent memory interfaces. The physical connection between the CPU and the FPGA is created using high-speed connectivity.

The shared memory structure between the CPU and the FPGA is depicted in Figure 6.12. For FPGA, the DRAM access granularity is cache line. The FPGA, on the other hand, is unable to access the DRAM directly [15]. To access DRAM data, it must, instead, submit read/ write requests to the coherent cache system. Figure 6.12 shows how the FPGA and CPU share the last level cache on the CPU. The shared memory allows the FPGA and CPU to communicate data efficiently.

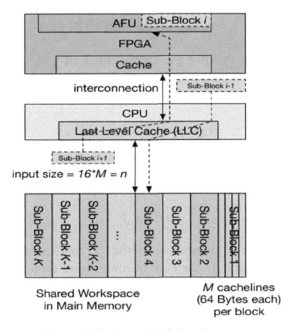

Figure 6.12 A model of shared memory.

6.3 Embedded Systems Applications in Biomedical Engineering: Case of the Pre-treatment of ECG Signal

6.3.1 The proposed techniques for embedded systems implementations

6.3.1.1 ADTF technique

The adaptive dual thresholding filter (ADTF) is a nonlinear filter proposed for the preprocessing of the ECG signal [16, 17]. The theoretical bases of this filter were inspired by the double thresholding median method published by Gupta, V. et al [18] used in image processing. In the case of the ECG signal, this filter aims to calculate three elements for each sliding window of the ECG signal: The mean of the sliding window, the upper threshold level, and the lower threshold level. The equation for a window means is as follows in equation (1):

$$g = \frac{1}{m} \sum_{i=n}^{n+m} \psi(i) \qquad (6.1)$$

"g" is the mean of the chosen window, "m" is the size of the window, and "$\psi(i)$" is the noisy ECG signal. The upper threshold equation is presented by (2):

$$Ht = g + \left[(Mx - g) \times \beta \right] \tag{6.2}$$

"*Ht*" is the upper threshold of the chosen window, "*Mx*" is the maximum value of the chosen window, and "β" is the thresholding coefficient. For the lower threshold, the equation is presented by (3):

$$Lt = g - \left[(g - Mi) \times \beta \right] \tag{6.3}$$

"*Lt*" is the lower threshold of the selected window, "*Mi*" is the minimum value of the selected window. "β" is the thresholding coefficient with:

$$0 < \beta < 1$$

As presented in [16, 17], lower values of the "β" coefficient are proposed for high noise concentration. In the case of lower noise concentration, a larger tolerance is required, i.e. higher values of β are recommended. The ADTF algorithm is the following for a given window of size [i;i+m]; with "φ" is the corrected ECG signal:

- If $\psi(i + m/2) > Ht \Rightarrow \varphi(i + m/2) = Ht$

- If $\psi(i + m/2) < Lt \Rightarrow \varphi(i + m/2) = Lt$

- If $Lt < \psi(i + m/2) < Ht \Rightarrow \varphi(i + m/2) = \psi(i + m/2)$

Figures 6.13 and 6.14 present two radical cases of the influence of noise on the morphologies of the ECG signal. For the first case, that of figure 6.13, the ADTF technique makes it possible to establish the different characteristics of the waves deteriorated by a very high level of noise of 10dB. This allows a correct analysis of the ECG signal waves (Ex: The P or T waves) despite a significant deterioration of their physiological characteristics. Regarding figure 6.14, the ECG signal presents the case of hyperkalemia with a critical deterioration of the major characteristics for the diagnosis of this case (Ex: The P wave, the QRS complex) by adding a very high level of noise of 10dB. The ADTF has also made it possible this time to restore these characteristics and to offer a signal ready for diagnosis.

For quantitative evaluation of the ADTF method, the use of the SNR improvement parameter is proposed. The SNRimp equation is presented by (6.4):

$$SNRimp = 10 \times \log_{10} \frac{\sum_{i=1}^{i=n} \left[\phi(i) - \psi(i) \right]^2}{\sum_{i=1}^{i=n} \left[\phi(i) - \varphi(i) \right]^2} \tag{6.4}$$

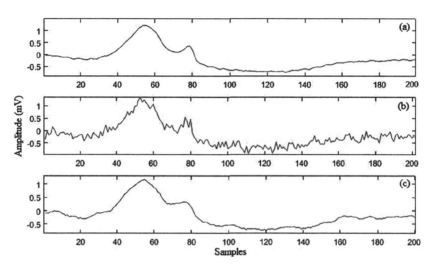

Figure 6.13 ECG signal denoising based on ADTF: the case of signal n°221 at 10 dB of WGN. (a) The original signal, (b) noise infected signal, (c) corrected signal.

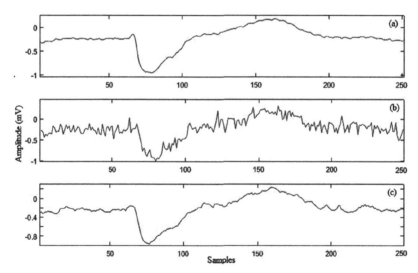

Figure 6.14 ECG signal denoising based on ADTF: the case of signal n°210 at 10 dB of WGN. (a) The original signal, (b) noise infected signal, and (c) corrected signal.

Figure 6.15 presents the results of the improvement of the signal-to-noise ratio (SNRimp) between the original signal, the noisy one, and the corrected signal. The same signals and noise levels showcased in the previous figure are respected. The SNRimp parameter varies between 8.15 for 0 dB of noise for the various signals analyzed and 2.59 for 20 dB of noise.

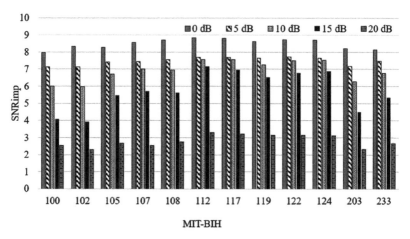

Figure 6.15   Evaluation of the SNRimp parameter for the ECG signal denoising based on ADTF.

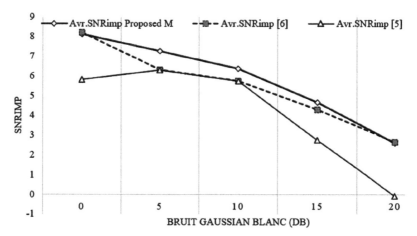

Figure 6.16   Comparative diagram of the evolution of the SNRimp parameter for the ECG signal denoising based on ADTF, EEMD-FR, and EEMD-GA.

Figure 6.16 presents a comparison diagram of the SNRimp parameter between the proposed technique and those of EEMD-FR [19] and EEMD-GA [20]. As shown in this figure, the ADTF offers competitive results to the EMD-GA technique [20] with much better results for a noise level greater than 0dB. Regarding the EEMD-FR technique [19], the ADTF offers much better results for all the noise levels presented in this study.

6.3.1.2 DWT Technique

The discrete wavelet transform (DWT) is a signal processing mathematical approach that is widely employed. The goal of this transformation is to use high-pass and low-pass filters to decompose a signal into multiple resolutions. Several high-pass "h" and low-pass "g" coefficients (Symlets coefficients, Debauchies coefficients, Coiflets) have been produced for a wide range of dilations and translations to obtain various types of signal analysis [21]. Equations (6.5) and (6.6) are used for a level of decomposition:

$$A[n] = \sum_{k=0}^{m} x[k] \times h[n-k] \qquad (6.5)$$

$$D[n] = \sum_{k=0}^{m} x[k] \times g[n-k] \qquad (6.6)$$

With A[n] being the result of the high-pass filter of signal x, this result is referred to as an approximation of this signal. D[n] is the result of the low pass filter of signal x, this result is referred to as a detail of this signal. m represents the size of the signal x or the number of samples of this signal [21]. Figure 6.17 presents an example of a decomposition of the signal x. Table 1 presents an example of the coefficients of the mother function of debauchees 2.

6.3.1.3 Hybrid DWT-ADTF technique
This section provides a description of an efficient method for denoising the ECG signal using an adaptive double thresholding filter (ADTF) and discrete

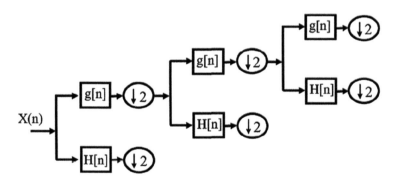

Figure 6.17 Example of a decomposition of the signal x.

Table 6.1 Coefficients of the function debauchies 2 (Db2).

Coefficients passe-haut		Coefficients passe-bas	
h [0]	−0,12940952	g [0]	−0,48296291
h [1]	0,224143868	g [1]	0,836516304
h [2]	0,836516304	g [2]	−0,22414387
h [3]	0,482962913	g [3]	−0,12940952

wavelet transform (DWT) [17, 22]. The purpose of this method is to combine the advantages of the two methods to improve the denoising of the ECG signal. The purpose of the proposed ADTF-DWT hybrid method is to deal with EMG electromyogram noises, power line frequency interference (50Hz), and high-frequency noises that could disturb the ECG signal.

The proposed algorithm is based on three denoising steps. This process makes it possible to successively reduce the noise of the ECG signal:

STEP 1: The DWT decomposes the ECG signal into various frequency bands. The Debauchies coefficients 6 (dB6) are the wavelet coefficients utilized in this approach. These coefficients demonstrate the best results in this technique when compared to others. This is owing to db6's resemblance to various ECG signal morphologies. Details 1 and 2 (D1 and D2), as illustrated in Figure 6.18, concentrate on a major portion of the noise in the ECG signal [23]. In the case of the ADTF-DWT technique, we propose that these details be removed in the first stage.

STEP 2: The suggested method for the second step is based on applying ADTF to the corrected signal from the first stage. The filtering window for this stage is 10 samples of the ECG signal, and the β parameter deployed for this step is 10%. For this stage, these options produce better results. Table 6.2 illustrates the impact

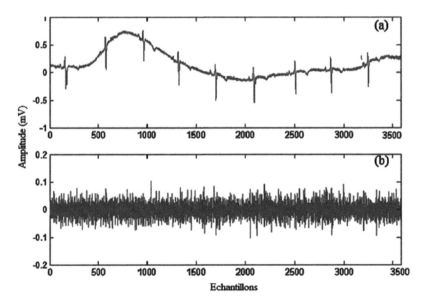

Figure 6.18   Step 1 of the proposed hybrid method. (a) original signal, (b) D1 signal + D2 signal.

Table 6.2 The influence of the β parameter on the results of the proposed hybrid filter.

	MIT-BIH	**5%**	**10%**	**15%**	**20%**
SNRimp	101	6.82	8.69	7.54	7.16
SNRimp	115	8.72	9.20	8.92	8.60

of the β parameter on the outcomes of the proposed hybrid filter using the SNRimp parameter and a white Gaussian noise level of 10dB applied to the original signal.

STEP 3: The purpose of the last step of this technique is to include a correction stage for the highest peaks of the ECG signal. These peaks present significant waves of the analyzed signal (the peaks of the R and S waves). This step aims at comparing the samples with a thresholding level identical to 70% of the maximum absolute value of the signal of the output of step 2. If a sample, at its absolute value, exceeds the thresholding level, the final value of the corrected signal is that of the output of step 1 and not that of step 2, otherwise, the output remains that of the latter.

Figure 6.19 presents the results of WGN denoising with a white noise level of 15dB. As shown in this figure, the proposed hybrid method provides an important solution to deal with noises that can disturb the ECG signal. The qualitative results of filtering the ECG signal based on the proposed method made it possible to see the good performance of this method. To further judge these results, a quantitative study will make it possible to study and compare the statistical results of this method with other research work presented recently in the state of art.

Table 6.3 provides a comparison of the results of the SNRimp parameter for the case of a 5 dB WGN noise for a variety of signals from the MIT-BIH database proposed by [24]. In this statistical study, it is clear that the hybrid ADTF-DWT method offers quite competitive and better results than those proposed by the ADWT method [25] for the different signals proposed in this study.

Table 6.4 shows a statistical comparison between the ADTF method and the hybrid ADTF-DWT method. The statistical parameter studied, in this case, is the percentage root-mean-square difference (PRD) parameter for a WGN noise with a signal-to-noise ratio of 5dB. The equation of the PRD parameter is presented by (7):

$$PRD(\%) = 100 \times \frac{\sum_{i=1}^{i=n}\left[\phi(i) - \varphi(i)\right]^2}{\sum_{i=1}^{i=n}\left[\phi(i)\right]^2} \qquad (6.7)$$

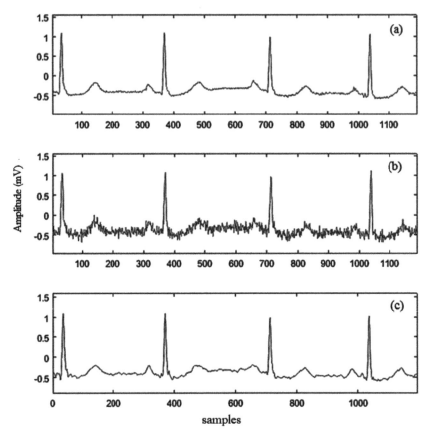

Figure 6.19   Filtering of high-frequency noise: the case of the MIT-BIH 101 signal with 15 B of the WGN. (a) the original signal, (b) noise infected signal, (c) corrected signal.

Table 6.3   The comparison of filtering results of WGN interference of 5dB noise level based on SNRimp parameter in some of the signals from MIT-BIH database.

MIT-BIH	ADWT	ADTF-DWT
100	9.40	9.70
101	9.09	10.23
103	7.13	9.10
113	7.82	9.33
115	7.19	9.45
117	8.62	9.34
119	7.27	8.13
122	7.86	8.07

Table 6.4 The results comparison of the WGN denoising of a noise level of 5 dB based on the PRD parameter in some of the signals of the MIT-BIH database.

MIT-BIH	ADTF	ADTF-DWT
100	24.55	18.26
103	25.23	19.61
105	24.53	20.56
115	24.74	19.46

As shown in this table, the hybrid method shows better results than the ADTF. This is due to the addition of the frequency analysis offered by the DWT which offers a more precise analysis for the denoising phase of the ECG signal.

6.3.2 Implementation of the ADTF technique using multi-CPU architectures

In this section, the ADTF algorithm for ECG signal filtering is evaluated using several embedded architectures, including a Raspberry 3B+ and an Odroid XU4. To take advantage of the parallelism in the architectures chosen, the implementation is based on C/C++ and OpenMP [26]. Different ECG signals proposed in the MIT-BIH Arrhythmia database with a sampling frequency of 360 Hz were used to validate the evaluation.

Mean Square Errors (MSE) and Signal to Noise Ratio (SNR) is computed for 12 ECG records of 10s to evaluate the denoising performance of the ADTF algorithm's C/C++ code compared to the Matlab code. In terms of excellent SNR with reduced MSE, the experimental results in Figures 6.20 and 6.21 reveal that the C/C++ program delivers more accurate denoising than the Matlab code.

A robust real-time implementation is required after demonstrating the algorithm's noise reduction performance. The results of the C/C++ implementations are better than those of the Matlab implementation. However, they are still a long way from real-time, as shown in Figure 3.10, with 1.56 milliseconds for Desktop, 9.2 milliseconds for Raspberry, and 6.8 milliseconds for XU4; while the real-time constraint is 2.77 ms imposed by the acquisition frequency.

OpenMP parallel programming is utilized to optimize the specified execution times and it provides great results compared to the naive C/C++ implementation. The ECG signal is divided among threads. Each thread runs the program on a subset rather than the entire signal. The processing time is depicted in Figure 3.11. To process one sample, the Raspberry architecture

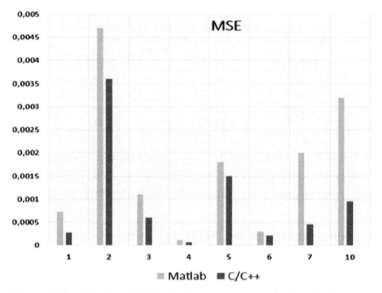

Figure 6.20   Matlab and C/C++ MSE comparison of the denoised signals.

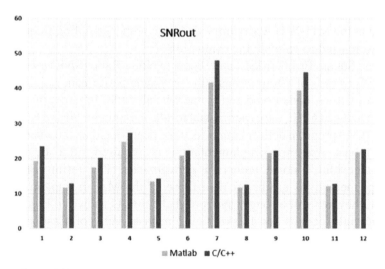

Figure 6.21   Matlab and C/C++ MSE comparison of the denoised signals.

takes an average of 7.5 milliseconds, the XU4 architecture takes 2.34 milliseconds, and the desktop takes 0.34 milliseconds.

The results allowed us to exclude raspberry as a solution due to the processing time exceeding 2.77 milliseconds. Despite its low energy consumption and weight, this architecture can't process the algorithm in real-time. The

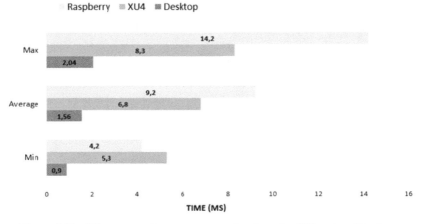

Figure 6.22 Min, max, and average processing times by different architectures.

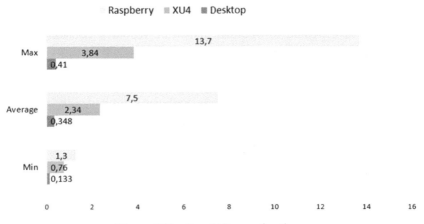

Figure 6.23 OpenMP executing time.

desktop had an extremely fast processing time of 0.34 milliseconds, demonstrating great performance. However, the high-power consumption means that this system does not match the dependability requirements. The XU4 architecture, on the other hand, met the time constraint, making it the ideal choice for this case. Furthermore, its low consumption in terms of power and lightweight supports the decision.

6.3.3 HLS implementation of ADTF technique using FPGA architecture

In this part, we present an HLS architecture of the ADTF algorithm dedicated to the implementation of FPGA. The implementation of FPGA of this

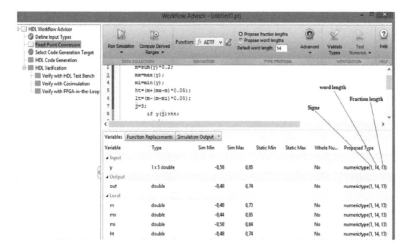

Figure 6.24   Automatic conversion of Matlab code to VHDL code.

algorithm is investigated to optimize hardware resource consumption and meet real-time processing requirements.

Today's knowledge of new technologies paves the way for the integration of millions of transistors on a single chip at great density. The development of a new level of design based on behavioral synthesis is limited to the last generations of FPGAs and requires starting with the behavioral specifications of an algorithm. For example, because FPGA technology has limited resources, we must regulate data word lengths to attain performance requirements and make optimum use of resources. However, it is too complicated when we need to transform algorithms written in double-precision floating-point to fixed-point implementations while maintaining mathematical accuracy. These tasks can be made easier by using Matlab's HDL coder [27]. The fixed-point conversion procedure is automated and managed by the tool. It proposes word lengths and precision criteria for all operations in the design using simulation data from a test bench file. These parameters can be used exactly as they are, or we can override them with our own. The example of automatic conversion of Matlab code to VHDL code is shown in figure 6.24.

Quartus II software was used to test and confirm the resulting code. Then, the ModelSim ALTERA is utilized to ensure that the produced VHDL description behaves correctly.

Quartus II software is used to generate the architecture's Register Transfer Level or (RTL). The generated architecture's input data is a five-element window. While ECG signal requires real-time processing, the signal must be handled as soon as it is obtained. The signal must be processed at

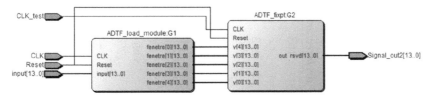

Figure 6.25 The RTL of the ADTF architecture.

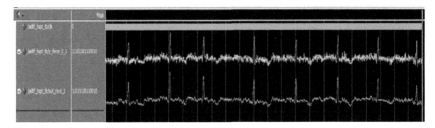

Figure 6.26 Simulation results of the ADTF architecture applied to signal No. 100 of MIT-BIH database with 10 dB WGN.

the same rate as it is acquired to be considered real-time. So, as shown in Figure 6.25, we added a block specifying a shift register with five elements with the same frequency as the sampling frequency (Based on the recorded signals from the MIT-BIH database, the frequency is 360 Hz), allowing online signal processing [28].

Figure 6.25 depicts the global architecture with the following inputs: input signal (14 bits), reset input, and the two clocks: Clk to load input data and Clk-test for the ADTF fixpt-module. The CLK-test period should be significantly shorter than the CLK period to ensure real-time processing.

ModelSim ALTERA software was used to run the simulation revealing that the resultant VHDL program behaved well. Figure 6.26 shows the simulation results of the ADTF code applied to the MIT-BIH database signal number 100 to which we added 10 dB of white Gaussian noise, allowing us to see the architecture's response in real-time due to the test clock, which enables the processing before the arrival of the next sample as input to the architecture.

The processing time is about 0.11 ms operating at 3.6 kHz when the input signal is sampled at 360 Hz, which means the time between two samples is 2.77 ms. As a result, the processing carried out by the architecture largely responds to the real-time constraint.

We compare the synthesis results of the ADTF architecture to the architecture proposed in [29] for the same algorithm in terms of resource use of different Intel ALTERA FPGA devices in Table 6.5.

Table 6.5 Comparison of the resources used by the optimized ADTF and the ADTF architecture proposed in [29].

	Arria II GX		Cyclone IV E		Cyclone IV GX		Cyclone V	
	ADTF [29]	Optimized architecture	ADTF [29]	Optimized architecture	ADTF [29]	Optimized architecture	ADTF [29]	Optimized architecture
Total Registers	245	154	244	154	245 (2%)	154 (1%)	251	163
Total logic elements	4%	2%	1590 (25%)	624 (10%)	1754 (17%)	812 (6%)	662 (1%)	253 (<1%)
Total pins %	17%	18%	33%	31 (34%)	37%	31 (38%)	11%	31 (12%)
Total block memory bits	0 %	0%	0 %	0 %	0 %	0 %	0 %	0%
DSP blocs	4 blocs (2 %)	2 blocs (<1%)	–	–	–	–	2 blocs (1 %)	3 blocs (2%)
Embedded Multiplier 9-bit elements	–	–	4 (13%)	2 (7%)	–	–	–	–

Table 6.5 illustrates that the comparison in terms of hardware resource consumption with the architecture proposed in [29] shows good optimization of our architecture, whether in terms of logical elements or registers.

As a result, the design spans extremely tiny areas in different FPGA targets, allowing us to free up resources for additional processing steps such as QRS extraction, arrhythmia detection, and more.

6.3.4 VHDL implementation of ADTF technique using FPGA architecture

In this chapter, we propose the description of a VHDL implementation of an ADTF architecture based on three modules: ALM, AFM, and ATM. The first module consists of loading cardiac data by ensuring real-time processing without using large memory space. The second module makes it possible to calculate the elements necessary for the treatment based on ADTF. The third module concerns the computation of threshold levels then the application of tests on ECG data and the assignment of output data. Figure 6.27 presents the RTL diagram of the ADTF architecture based on the three modules. In digital circuit design, the RTL is a design abstraction that models a digital circuit in terms of signals flow between hardware registers and the logic operations performed on these signals.

ADTF Load Module (ALM)

The aim of the ALM module, as shown in Figure 6.28, is to read entering data through the 11-bit Input signal port and prepare it for the modules that follow it. The operating principle of this module is based on a shift register with a processing frequency identical to that of the sampling frequency of the input ECG signal. In this case, it is the signals from the MIT-BIH database of 360 Hz.

The ADTF processing window in this study is 5 samples, so the size of the shift register is 5 elements. This approach allows the online processing of

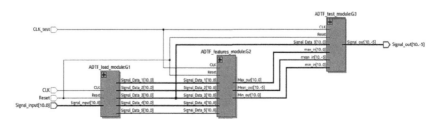

Figure 6.27 RTL diagram of the ADTF total architecture.

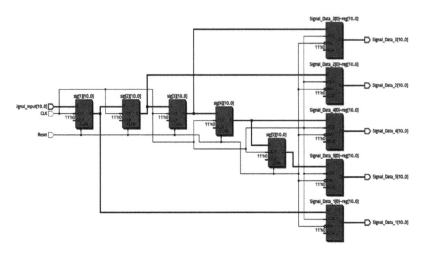

Figure 6.28   RTL diagram of ALM module.

cardiac data without occupying a large memory space for storing data before and after processing. Each item in the register is bound to an output of the same data size.

ADTF Features Module (AFM)

The AFM module, illustrated in figure 6.29, prepares the data necessary for processing, especially, the average of the ADTF window, the maximum, and the minimum of the window. The analyzed data comes from the output of the ALM module. The processing frequency of this module is higher than that of the ALM module. This ensures real-time analysis of the ECG signal. We propose a frequency of 3600 Hz, which is 10 times the frequency of reading the input signal proposed for the ALM module.

The average calculation is based on 4 sum operations plus division by 5. The output data size of this operation is 30 bits, 15 bits for the sum operation, and 30 bits after the division. After this operation, we proposed to minimize the size of the output from the mean to 16 bits, 11 bits for the integer part of the output value, and 5 bits for the fixed-point part of this value. For the calculation of the maximum and the minimum, we proposed two test loops on the applied input data respectively. The output is 11 bits for both operations.

ADTF Test Module (ATM)

The principle of the ATM module, as shown in figure 6.30, is to compute the thresholding values of the two levels Ht and Lt. The data required for the computation comes from the MCP module, namely the average, the

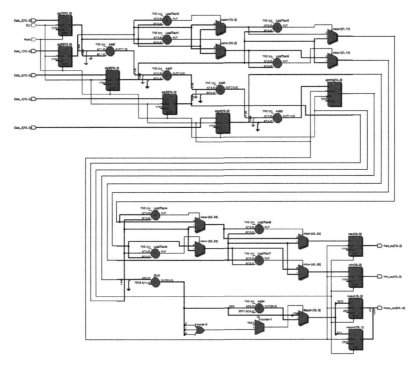

Figure 6.29   RTL diagram of AFM module.

minimum, and the maximum of the window. The calculation frequency of the ATM module is the same as the AFM module. The ATM module process is as follows:

- Calculation of the value of the parameter β: In this step, we suggest the creation of an 11-bit register to store the value β. The MSB is reserved for zero, which is the integer value of β hence β = 0.1. The bits that follow are the fixed-point binary values of this parameter.

- Calculation of the Ht and Lt values: For this stage of processing, the calculation begins with a subtraction operation (ex: (Mx-g) for the Ht), then a multiplication of the output to the value of β, and, in the end, subtraction or addition to the mean value for the calculation of Ht and Lt respectively.

- Tests and assignments: For the test phase, the comparison concerns the Ht and Lt values with the value of the median sample of the ADTF window. Each of the tests is only interested in the integer values Ht and Lt of 11 bits each. This is due to the ADTF's median window sample

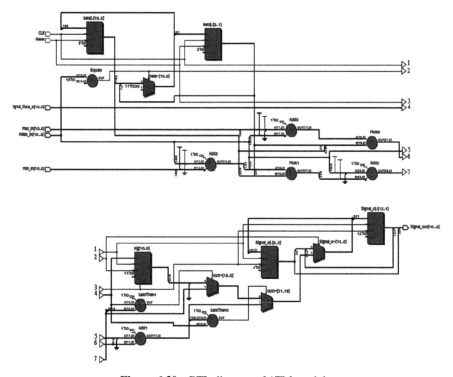

Figure 6.30   RTL diagram of ATM module.

size of 11 bits. The result of these tests is an assignment of a value to the output signal, whether it is those of Ht, Lt, or the median sample.

The size of an output data is 16 bits: 11 bits reserved for the integer part and 5 bits for the fixed point whether it is the value assigned to the output, namely Ht, Lt, or the median sample of the window of the ADTF. For the first two values, the output is an adjustment of their 16-bit values. For the case of assigning the ADTF's mid-window sample value to the output, which is 11 bits, the output is 11 bits of the integer value same as the size of this sample, and 5 bits of logical zeros for the fixed-point part.

In this study, the performance simulation of the proposed ADTF architecture is provided by the ModelSim software. The latter is an HDL simulation tool developed by Mentor Graphics for the simulation of hardware description languages such as VHDL, Verilog, and SystemC. ModelSim can be used in conjunction with Altera Quartus or Xilinx ISE or on its own. The simulation is performed manually using the graphical user interface (GUI) or automatically with scripts.

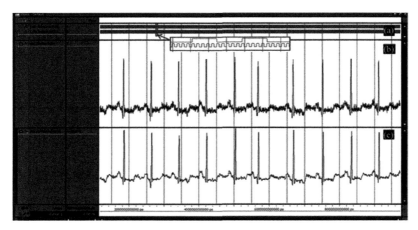

Figure 6.31 Real-time denoising under ModelSim: the case of signal n°100 at 20 dB of noise WGN. (a) CLK and Internal CLK clock signals, (b) input signal, and (c) output signal.

We propose a qualitative study of the results of the real-time simulation of the filtering of ECG signals from the MIT-BIH database. This study allows us to evaluate the proposed ADTF architecture performances on VHDL.

Figure 6.31 shows an example of real-time denoising of signal n°100 at 20dB of white Gaussian noise under ModelSim. This figure illustrates the flow of data according to the clock for real-time monitoring. As shown, the proposed ADTF architecture offers an efficient solution for high-frequency noise filtering in real-time.

Figure 6.32 presents a qualitative comparison between the results of the ADTF-based denoising under "Soft-ADTF" software and the real-time denoising based on the proposed VHDL architecture of the ADTF "VHDL-ADTF". This figure shows a case of adding 10dB of blue-colored noise to signal n°203. Colored noise is noise with varying integrated power in different frequency bands during the same period. It will have a distinct power spectrum depending on whether it is pink, blue, or another color. The noise density fluctuates at different frequencies depending on this power. What makes the different natures of colored noise closer to those actually found is that influences physiological signals for example. In this figure, a great similarity is observed between the results of the filtering of Soft-ADTF and VHDL-ADTF in real-time. This reflects the great performance of the proposed VHDL implementation.

Table 6.6 shows a comparison of the evaluation results of the SNRimp parameter for the denoising of colored noises for the case of signal n°203. The two noises proposed are blue noise and pink noise at two noise levels 5dB and 10dB.

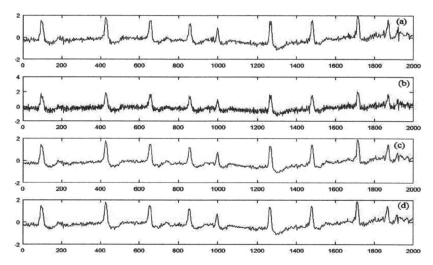

Figure 6.32   ECG signal denoising based on ADTF: the case of signal n°203 at 10 dB of blue-colored noise. (a) The original signal, (b) noise infected signal, (c) soft-ADTF corrected signal, (d) VHDL-ADTF corrected signal.

Table 6.6   Evaluation results of the SNRimp parameter for colored noise denoising: the case of signal n°203.

Methods	Colored noises			
	Blue (5 dB)	**Pink (5 dB)**	**Blue (10 dB)**	**Pink (10 dB)**
Soft - ADTF	9.37	1.43	6.91	1.18
VHDL - ADTF	9.54	1.39	7.98	1.22

In Table 6.6, it is clear that the statistical results based on the SNRimp parameter present a very small difference in this case in favor of the VHDL implementation of the ADTF with real-time processing of the ECG signal. This approves the high performance of the proposed implementation. In the case of WGN noises, the statistical results based on the SNRimp parameter present very small differences in favor of the Soft-ADTF.

These small differences in the statistical results do not reflect an improvement or a change in the general course of treatment with ADTF. They are related to the size of the data during processing as well as the choice of the fixed point of 5 bits for the fractional part. Soft-ADTF is built on a 64-bit machine with floating-point processing. But in the case of VHDL-ADTF, we have proposed fixed-point based processing to reduce the complexity of the processing as well as to further simplify the proposed implementation, which allows it to be implemented in different FPGA devices.

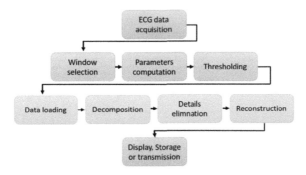

Figure 6.33 The algorithm block diagram.

6.3.5 VHDL implementation of hybrid DWT-ADTF technique using FPGA architecture

In this section, the work's goal is the on-board implementation of the hybrid technique based on ADTF and DWT for ECG denoising. The two filters, ADTF and DWT, are developed in VHDL (VHSIC Hardware Description Language) for an FPGA implementation using the Quartus II tool and the Modelsim simulation environment.

The hybrid approach combines ADTF and DWT allowing the noise in the ECG signal to be reduced successively [17]. The entire procedure is depicted in figure 6.33, which shows the ECG signal going through two rounds of noise reduction:

The ADTF is applied to the noisy signal in the first step of this approach; the window size is set to 10 samples and the thresholding coefficient is set to 0.1 (10 %).

The second step is to apply DWT to the rectified signal from the previous stage, which decomposes the signal into numerous frequency bands. Debauchies dB4 is the wavelet mother utilized in this situation. The coefficients of this wavelet are the most comparable to the ECG signal in terms of similarity. Because details of levels 1 and 2 (D1 and D2) concentrate a significant amount of noise after decomposition, we decided to remove them. The denoised signal is then processed using the inverse DWT.

Because the design is intended to be implemented on various FPGA boards, it is based on a structural description divided into blocks. The different blocks define the ADTF-DWT modules independently to allow for parallel processing reducing processing time [30]. The suggested method's architecture comprises two main blocks: one for the ADTF denoising step and the other for the DWT denoising step. Figure 6.34 displays the global architecture's RTL schema.

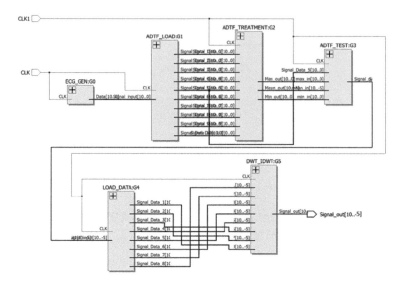

Figure 6.34   The RTL of the hybrid technique.

The ADTF block consists of three functional blocks: ADTF-LOAD (FB1) or shift register that prepares the window for the second functional block, ADTF-TREATMENT (FB2), which determines the ADTF process's required parameters, and the third functional block, ADTF-TEST (FB3), which performs a thresholding operation on the window's median value.

The output of the first block passes via the second block where the DATALOAD functional block (FB4) prepares an eight-element window. The DWT-IDW functional block (FB5) then applies the DWT, eliminates the noisy details, and applies the inverse DWT to this portion of the signal.

MIT-BIH Physionet's Arrhythmia database is used the test the VHDL architecture's functionality. The signals are sampled at a frequency of 360 Hz with a resolution of 11 bits. Before the denoising procedure, White Gaussian Noise (WGN) is mixed with the original signals for the test.

The simulation results of the architecture applied to signal number 100 of the MIT-BIH database with a White Gaussian Noise of 20 dB are shown in Figure 6.35. The simulation results show that the method performs well in noise reduction without distorting the original signal preserving, thus, its morphology. After timing verification, the system replies in 0.3 milliseconds using a 50 kHz processing clock, mainly answering the real-time constraint posed by the acquisition frequency (360 Hz).

Figures 6.36, 6.37, 6.38, and 6.39 show the hardware resources consumption by the hybrid architecture on FPGA INTEL-ALTERA boards.

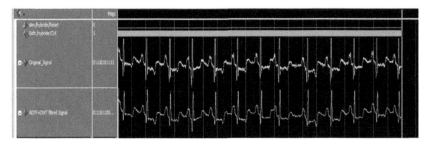

Figure 6.35 Simulation result of architecture applied to the signal number100 of the MIT-BIH database.

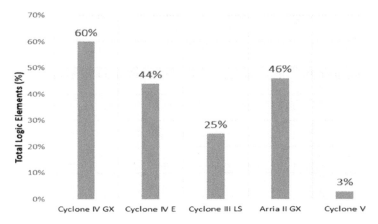

Figure 6.36 The architecture's total logic elements in different FPGA targets.

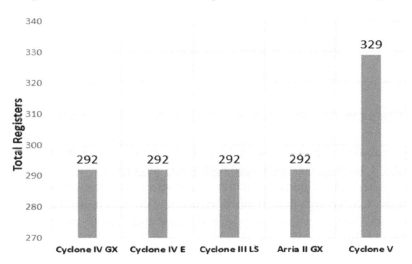

Figure 6.37 The architecture's total registers in different FPGA targets.

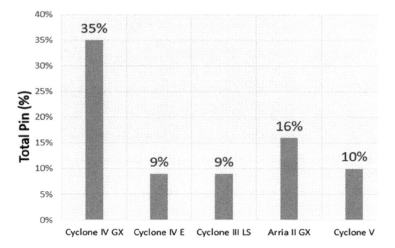

Figure 6.38   The architecture's total pins in different FPGA targets.

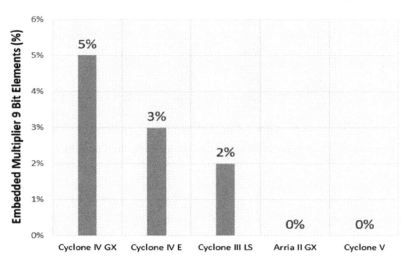

Figure 6.39   The architecture's embedded multipliers in different FPGA targets.

They present the consumption in terms of the total logic elements, utilized registers, total pins, and embedded multipliers employed.

The hybrid technique's architecture doesn't require expensive FPGA boards to achieve high performance. Hence, the devices utilized in the comparison are categorized as low-power and low-cost. The Cyclone and Arria families are studied.

With a total of 292 registers for Cyclone IV GX, Cyclone III LS, Cyclone IV E, and Arria II GX, and 329 for Cyclone V. The hybrid architecture consumes less than 1% of the total registers for all FPGA devices. The occupancy of logic elements in Cyclone V and Cyclone IV GX varies between 3% and 60%, respectively.

The global architecture has 28 pins with 11 pins for the input signal, which is coded in 11 bits, 16 pins for the output or denoised signal, and one pin for the clock, which has a percentage of 9% for Cyclone IV-E and Cyclone III-LS, 10% for Cyclone V, 16% for Arria II-GX, and 35% for Cyclone IV-GX.

Some devices use an integrated multiplier of 9-bit elements to optimize multiplications. So, the architecture requires eight embedded multiplier 9-bit elements, which account for 5% of the Cyclone IV-GX, 3% of the Cyclone IV-E, and 2% of the Cyclone III-LS.

To validate the proposed architecture, we implemented the design on the Intel-Altera FPGA-DE1 board. Based on the test signals, we also developed a study on the state of the architecture's output signals. The SignalTap II tool included in the Quartus II embedded architecture design program performs the validation step. The results achieved using ModelSim are confirmed by the architecture's implementation on the FPGA-DE1 board as can be seen in figure 6.40.

6.4 Conclusions

Proposing algorithms or merging between them always requires behavior algorithm validation. This validation on conventional machines such as workstations does not imply the implementation in embedded architecture. Even if homogeneous (CPU/CPU) or heterogeneous (CPU-GPU/ CPU-FPGA/ CPU-DSP), this architecture requires a detailed study of the software and hardware parts. This study includes temporal and algorithmic complexity as well as the study of the target architecture. Furthermore, the implementation of ECG signal preprocessing algorithms in embedded architecture gives flexibility in terms of temporal constraints. This flexibility is based on optimizing the target architecture to propose implementations that meet the different real-time requirements.

These requirements in the medical field make the implementation task difficult if the embedded architecture does not give the necessary flexibility to optimize the consumed resources, which will influence the processing time directly. Through this work, we have proposed a study on the two filtering algorithms, DWT and ADTF. The use of these two filtering techniques in a single algorithm to exploit the strengths of ADTF and DWT. The evaluation of the algorithm has shown that the results are better than ADTF, which

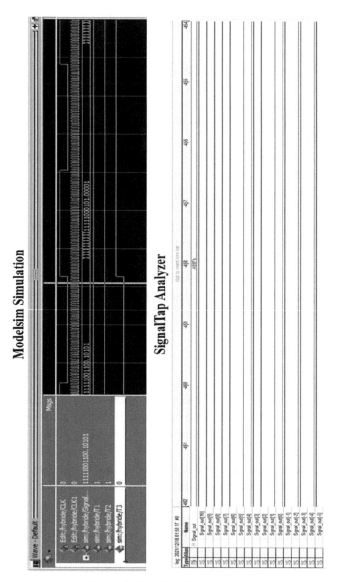

Figure 6.40　SignalTap validation.

makes this technique efficient. We have also proposed implementations in different architectures. These implementations satisfied the time constraints, but the design of the algorithm architecture in the FPGA board showed that this type of architecture showed more favorable results than the other. Either at the level of adequate architecture with the algorithm, processing time as well as the low cost and energy consumption. Still, the problem with this type of architecture compared to the CPU-FPGA system is the complexity of code development, which makes this architecture limited. Another alteration that has been proposed in this work is the use of code generators offered by Matlab. This type of tool is very effective if we do not take into account the optimization of the target architecture, which poses a significant limitation. Besides, the use of multi-core can accelerate the processing, but the problem here is the time taken in scheduling between different processors.

As a solution, the use of architecture based on multi-core and FPGA. This solution will allow us to exploit the advantages of FPGA and multi-core. But we need to use high-level languages like OpenCL and OpenMP to use the features of FPGA for the acceleration of the algorithm parts as well as the multicore. However, this exploitation requires a detailed study of the software and hardware based on the H/S Mapping approach to have an optimal implementation that satisfies the different environmental and algorithmic constraints. Therefore, in future work, a heterogenous architecture based on multi-core and FPGA will be used. Other processing stages will be added to the implementations to realize a real-time monitoring system of patients' cardiac status.

References

[1] S. Heath, "Embedded systems design," *Elsevier*, 2002.

[2] W. Jenkal, «Conception et implémentation des algorithmes d'analyse des signaux cliniques ecg : Vers un système embarqué,» *PhD thesis National School of Applied Sciences*, 2018.

[3] S. Youngsoo, K. Choi, and T. Sakurai, "Power optimization of real-time embedded systems on variable speed processors," in *IEEE/ACM International Conference on Computer-Aided Design. ICCAD-2000*, 2000.

[4] H. Ghasemzadeh, S. Ostadabbas, E. Guenterberg, and A. Pantelopoulos, "Wireless medical-embedded systems: A review of signal-processing techniques for classification," *IEEE Sensors Journal*, vol. 13, no. 2, pp. 423–437, 2013.

[5] R. M. Rangayyan, "Biomedical signal analysis," *John Wiley & Sons*, vol. 33, 2015.

[6] D. Etiemble, "45-year CPU evolution: one law and two equations," *arXiv preprint arXiv:1803.00254*, 2018.

[7] N. Harki, A. Ahmed, and L. Haji, " CPU scheduling techniques: A review on novel approaches strategy and performance assessment.,"

Journal of Applied Science and Technology Trends , vol. 1.2, pp. 48-55, 2020.

[8] I. ZAGAN, " Improving the performance of CPU architectures by reducing the Operating System overhead," in *IEEE 3rd Workshop on Advances in Information, Electronic and Electrical Engineering (AIEEE). IEEE*, 2015.

[9] S. W. Smith, "The Scientist and Engineer's Guide to Digital Signal Processing," *California Technical Publishing*, pp. 503-534, 1997.

[10] D. Marković and R.W. Brodersen, "DSP architecture design essentials, " *Springer Science & Business Media*, 2012.

[11] P. Eles, K. Kuchcinski and Z. Peng, "System synthesis with VHDL," *Springer Science & Business Media*, 2013.

[12] C. H. Roth. and L. K. John, "Digital systems design using VHDL," *Cengage Learning*, 2016.

[13] F. Vahid, "Digital Design with RTL Design, Verilog and VHDL," *John Wiley & Sons*, 2010.

[14] D. Iorga, A. F. Donaldson, T. Sorensen and J. Wickerson, " The Semantics of Shared Memory in Intel CPU/FPGA Systems," in *Proc. ACM Program. Lang. 5, OOPSLA*, 2021.

[15] C. Zhang, R. Chen and V. Prasanna, "High throughput large scale sorting on a CPU-FPGA heterogeneous platform," in *nternational Parallel and Distributed Processing Symposium Workshops (IPDPSW). IEEE*, 2016.

[16] W.Jenkal, R. Latif,A. Toumanari, A. Dliou and O. El B'charri, "An efficient method of ECG signals denoising based on an adaptive algorithm using mean filter and an adaptive dual threshold filter," *International Review on Computers and Software (IRECOS),* Vols. 10(11), pp. 1089-1095, 2015.

[17] W.Jenkal, R. Latif,A. Toumanari, A. Dliou, o. El B'charri and F. M. Maoulainine, "An efficient algorithm of ECG signal denoising using the adaptive dual threshold filter and the discrete wavelet transform," *Biocybernetics and Biomedical Engineering,* vol. 36(3), pp. 499–508, 2016.

[18] V. Gupta, V. Chaurasia and M. Shandilya, " Random-valued impulse noise removal using adaptive dual threshold median filter," *Journal of visual communication and image representation,* vol. 26, pp. 296–304, 2015.

[19] K. M. Chang, " Arrhythmia ECG noise reduction by ensemble empirical mode decomposition," *Sensors,* vol. 10(6), pp. 6063-6080, 2010.

[20] P. Nguyen and J.M. Kim, " Adaptive ECG denoising using genetic algorithm-based thresholding and ensemble empirical mode decomposition," *Information Sciences,* vol. 373, pp. 499–511, 2016.

[21] H. Khorrami and M. Moavenian, "A comparative study of DWT, CWT and DCT transformations in ECG arrhythmias classification," *Expert systems with Applications,* vol. 37(8), pp. 5751–5757, 2010.

[22] W. Jenkal, R. Latif, A. Toumanari, A. Elouardi, A. Hatim and O. El Bcharri, "Enhancement and compression of the electrocardiogram signal using the discrete wavelet transform," in *International Conference on Wireless Technologies Embedded and Intelligent Systems (WITS),* 2017.

[23] S. Banerjee, R. Gupta and M. Mitra, "Delineation of ECG characteristic features using multiresolution wavelet analysis method," *Measurement,* vol. 45(3), pp. 474–487, 2012.

[24] G.B. Moody and R.G. Mark, "The impact of the MIT-BIH arrhythmia database," *IEEE Engineering in Medicine and Biology Magazine,* vol. 20(3), pp. 45–50, 2001.

[25] M.A. Awal, S. S.Mostafa, M. Ahmad and M. A. Rashid, "An adaptive level dependent wavelet thresholding for ECG denoising," *Biocybernetics and Biomedical Engineering,* vol. 34(4), pp. 238–249, 2014.

[26] S. Mejhoudi, R. Latif, A. Saddik, W. Jenkal and A. El Ouardi, "Speeding up an Adaptive Filter based ECG Signal Pre-processing on Embedded Architectures," *International Journal of Advanced Computer Science and Applications (IJACSA),* vol. 12, no. 5, 2021.

[27] T. Bonny, "Chaotic or hyper-chaotic oscillator? Numerical solution, circuit design, MATLAB HDL-coder implementation, VHDL code, security analysis, and FPGA realization," *Circuits, Systems, and Signal Processing ,* vol. 40.3, pp. 1061–1088, 2021.

[28] S. Mejhoudi, R. Latif, W. Jenkal and A. El Ouardi, "Real-time ECG Signal Denoising Using the ADTF Algorithm for Embedded Implementation on FPGAs," in *4th World Conference on Complex Systems (WCCS). IEEE,* 2019.

[29] W. Jenkal, R. Latif, A. Toumanari, A. Elouardi, A. Hatim and O. El Bcharri, "Real-Time Hardware Architecture of the Adaptive Dual Threshold Filter based ECG Signal Denoising," *Journal of Theoretical and Applied Information Technology,* vol. 96, no 14, pp. 4649–4659, 2018.

[30] S. Mejhoudi, R. Latif, W. Jenkal, A. Saddik and A. El Ouardi, "Hardware Architecture for Adaptive Dual Threshold Filter and Discrete Wavelet Transform based ECG Signal Denoising," *International Journal of Advanced Computer Science and applications (IJACSA),* vol. 12, no. 11, 2021.

7

Acquisition and Processing of Surface EMG Signal with an Embedded Compact RIO-based System

Abdelouahad Achmamad, Mohamed El Fezazi, Atman Jbari

Electronic systems sensors and Nano-biotechnologies, National Graduate School of Arts and Crafts (ENSAM), Mohammed V University in Rabat, Morocco
Abdelouahad.achmamad@um5r.net.ma, elfezazi.med@gmail.com, atman.jbari@um5.ac.ma

Abstract

Surface electromyography (sEMG) signal has been the subject of much research for many years now. The surface EMG signal provides a non-invasive possibility of studying muscular activities produced during periods of contraction and relaxation. However, commercial sEMG systems are not flexible enough and not scalable to contemporary needs. Therefore, this chapter deals with the implementation of the signal conditioning circuit, real-time acquisition, and feature extraction of the sEMG signal. The entire embedded system was developed in a user-friendly and flexible way. For the EMG signal conditioning circuit, National Instrument Educational Laboratory Virtual Instrumentation Suite (NI-ELVIS II+) was used. Further tasks were implemented on a Field Programmable Gate Array (FPGA) and a real-time processor with LabVIEW programming. The implementation results were presented on a LabVIEW-based graphic user interface (GUI). It can be concluded that the evaluated time and frequency features such as sEMG envelope, mean absolute value (MAV), root mean square (RMS), mean power frequency (MPF), and median frequency (MDF) can be employed as valuable vital signs monitor to assist doctors to perform their particular spots regarding the state of the muscle.

7.1 Introduction

Nowadays, biomedical monitoring systems have increased their popularity in modern healthcare and medicine. The monitoring of bio-signals is not only serves to understand the physiological changes produced in the body, but also a powerful way to anticipate the occurrence of specific dysfunctions. One of the most widely used bio-signal types is surface electromyography (sEMG) signal, which provides the means to understand how skeletal muscle works. In fact, the sEMG signal expresses the electrophysiology of muscle contraction/relaxation, and it can be easily recorded with patch electrodes. Positive and negative electrodes are placed over the target muscle, while the third reference electrode has to be placed away from it [1, 2]. For the standard sEMG signal, the range of frequency content is typically from 20 to 2000Hz and its amplitude range is between 0 and 10 mV, depending on the type of investigation (non-invasive or invasive) [3, 4]. This non-invasive signal, therefore, contains abundant physiological insights that can be used for analysis in the studies such as muscle fatigue [5], strength [6], gait activity [7], and gesture [8], among other applications. On the other hand, the sEMG signal is inevitably affected by noises (e.g., electrode motion artifacts, muscle artifacts) that are generated from a variety of external sources during the recording [1]. These noises severely limit the utility of the sEMG signal, and thus the use of hardware efficient filters long before the sEMG processing step is crucial to exclude all the unwanted frequency components. Signal conditioning is a well-known electronic circuit. It usually consists of an analog conditioning stage that amplifies and filters the signal, followed by an analog-digital converter [4, 9, 10].

Several health-related electronic and software based on sEMG signal monitoring systems and signal processing can be found in the literature [11–14]. Despite the ongoing evolution of sEMG monitoring systems, many of them are still treadmill embedded systems because of their restricted characteristics. Most traditional designs suffer from various shortcomings. Similar to other systems, they perform specialized tasks, serve specific targets, or are suitable for certain technologies. Firstly, they lack a user-friendly graphical (GUI), and even if it is available, it is usually programmed by classical software, which is not easy for clinicians to develop their functional requirements. This makes the available commercial sEMG monitoring system hard to develop and reuse. Secondly, they do not provide pertinent information to medical experts to help them get the correct interpretation and medical diagnosis. Therefore, an urgent need to develop an embedded system that can overcome the aforementioned limitations. In this chapter, a comprehensive solution that includes a real-time sEMG monitoring system based on an embedded CompactRIO-9035 real-time controller and NI-ELVIS II+ board is proposed. In the first place, the EMG signal conditioning circuit was built using an instrumentation amplifier and analog high-pass filter cascaded with an analog low-pass filter, followed by the sEMG signal acquisition stage employing the FPGA integrated circuit. And more, the acquired sEMG signal has been processed on real time processor through CompactRIO-9035 real-time controller. The full chain processing algorithm was developed for the purpose of extracting the popular time and frequency features from each voluntary muscle contraction. In order to visualize the implementation results, considerable efforts have been carried out to create a new design of GUI based on LabVIEW software. The use of LabVIEW and data acquisition makes the real-time sEMG monitoring system more reliable, high-performance, and flexible. Consequently, the developed system is highly effective in those applications where it is required to monitor for a long time the muscle activity, and detect and distinguish voluntary muscle contractions.

On the whole, the study is summarized as follows:

- Pre-processing of sEMG signal.

- Acquiring the sEMG signal in real-time.

- Applying threshold-based algorithm to get only the information related to the contraction and relaxation phase of the skeletal muscle.

- Carrying out feature extraction for sEMG monitoring.

The remainder of the paper is organized as follows. Section 2 provides an EMG signal conditioning circuit. Section 3 presents the proposed process and signal processing-based methods. The implementation results are presented in section 4. Section 5 summarizes the conclusion.

7.2 EMG Signal Conditioning Circuit

7.2.1 Instrumentation amplifier

The instrumentation amplifier (IA) allows the differential amplification of the sEMG signal. Its objective is to measure the difference between two voltages. As illustrated in Figure 7.1, it is often made using three Op-Amp In-Amp. This configuration is characterized by a high input impedance and a high common mode rejection ratio (CMRR) provided in the first stage (differential amplifier). It offers a differential gain $A_d = 1 + \dfrac{R_2}{R_1}$. The second stage (difference

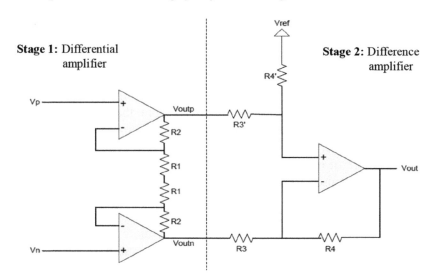

Figure 7.1   Classical three Op-Amp In-Amp.

amplifier) eliminates the common mode effect and provides a differential gain $A_d = \dfrac{R_4}{R_3}$ [15, 16].

The output voltages of the differential amplifier (stage 1) are given as:

$$V_{outp} - V_{outn} = \left(V_p - V_n\right) \times \left(1 + \frac{R_2}{R_1}\right) \qquad (7.1)$$

For the second stage (difference amplifier) we have:

$$V_{out} = \frac{R_4}{R_3} \times \left(V_{outp} - V_{outn}\right) \qquad (7.2)$$

Therefore, we obtain the Instrumentation Amplifier equation:

$$V_{out} = \frac{R_4}{R_3} \times \left(1 + \frac{R_2}{R_1}\right) \times \left(V_p - V_n\right) \qquad (7.3)$$

For the acquisition, we used the AD8226 circuit (see Figure 7.2) designed to use in medical instrumentation. AD8226 is a low-cost amplifier with high CMRR (more than 90 dB), operates on a supply range of signal voltages (single and dual supplies) that needs only one external resistor to set any gain

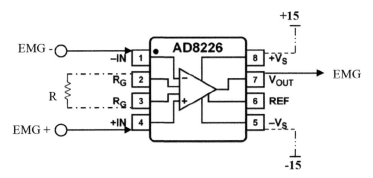

Figure 7.2 AD8226 circuit.

between 1 and 1000. This gain can be computed according to the following gain equation [17]:

$$G = 1 + \frac{49.4k}{R_G}$$ (7.4)

where, R_G is the gain-setting resistor.

7.2.2 Band pass filter

The useful information of the sEMG signal ranges from 20 to 2000 Hz. For this purpose, the recorded sEMG signal was filtered by cascading two filters: a high-pass filter with a cut-off frequency of ω_{c1}; a low-pass filter of , both filters constituting $[\omega_{c1}, \omega_{c2}]$ Butterworth band-pass filter of order 2. This filtering process removed the noise generated mainly by the motion artifact and the inherent instability of the signal [16].

The final design of the band-pass filter is given in Figure 7.3. Therefore, the transfer functions of the high-pass and the low-pass filters respectively can be given by:

$$T_1(\omega) = -\frac{\left(j\omega\sqrt{C_1 C_2 R_1 R_2}\right)^2}{1 + j\omega R_1\left(C_1 + C_2 + C_3\right) + \left(j\omega\sqrt{C_1 C_2 R_1 R_2}\right)^2}$$ (7.5)

where, high-pass filter cut-off frequency at -3dB is:

$$\omega_{c1} = \frac{1}{\sqrt{C_1 C_2 R_1 R_2}}$$ (7.6)

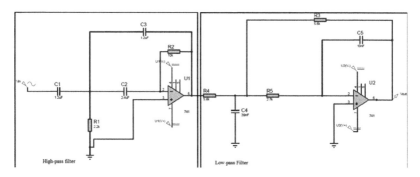

Figure 7.3   Band-pass filter circuit.

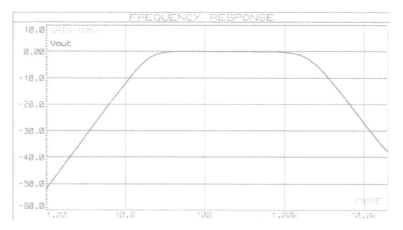

Figure 7.4   Band-pass filter frequency response.

$$T_2(\omega) = -\frac{\dfrac{R_3}{R_4}}{1 + j\omega R_5 R_3 C_5 \left(\dfrac{1}{R_3} + \dfrac{1}{R_4} + \dfrac{1}{R_5}\right) + \left(j\omega\sqrt{C_4 C_5 R_3 R_5}\right)^2} \qquad (7.7)$$

The low-pass filter cut-off frequency at −3dB is:

$$\omega_{c2} = \frac{1}{\sqrt{C_4 C_5 R_3 R_5}} \qquad (7.8)$$

The frequency response of the designed filter is illustrated in Figure 7.4:

7.2.3 Analog-to-digital converter

The Analog-to-Digital Converter (ADC) is used to convert the filtered sEMG signal into a digital signal for further processing in LabVIEW software. This important step is done using the ADC configuration of the NI-ELVIS board with a high resolution of 16-bits.

7.3 EMG Signal Processing

7.3.1 Flowchart description

Figure 7.5 clearly describes the full chain processing of the proposed flow-chart, where in the first part, the sEMG signal is read to then pass it via pre processing stage, whereby applying the threshold-based algorithm. In the pre processing stage, the sEMG signal is initially rectified to get only the real positive values. The full-wave rectifying is conducted using an absolute function with respect to Equation 7.9. After performing the full-wave rectification operation, the envelope of the sEMG signal is extracted. For the method using the moving average filter (MAF), Equation 7.10 is applied. This operation is carried out on the whole rectified sEMG signal via an analysis sliding window having 100 with a step number of one sample so that to make the linear envelope smoother. It is stated in the literature that N_w should fall between 10 *ms* and 150 *ms* for an accurate linear envelope [18]. Based on the amplitude variation of, the onset point can be highlighted by applying a threshold to . The threshold is a real positive value that limits the sEMG signal to be counted. This threshold is practically set at 14 µV.

$$sEMG_r = \begin{cases} sEMG & \text{if } sEMG \geq 0 \\ -sEMG & \text{if } sEMG < 0 \end{cases} \tag{7.9}$$

$$MAF(k) = \frac{1}{N_w} \sum_{i=0}^{N_w} sEMG_r(k+i) \tag{7.10}$$

where i = 1,2,…,N_w, N_w is the window length and is the number of step.

In the second part, the desired sEMG features are calculated for each detected contraction/relaxation cycle. Different indicators might be used to estimate the amplitude of sEMG. The root-mean-square (RMS) and mean average (MAV) are always adopted when quantifying the changes in muscle voluntary contraction. MAV and RMS parameters can be calculated by the following mathematical expressions, respectively [19, 20]:

$$MAV = \frac{1}{N}\sum_{n=1}^{N}|sEMG(n)| \tag{7.11}$$

$$RMS = \sqrt{\frac{1}{N}\sum_{n=1}^{N}sEMG(n)^2} \tag{7.12}$$

where n = 1,2,...,N and N is the length of sEMG signal.

In spectrum analysis, the sEMG signal is usually performed using a fast Fourier transform (FFT) to allow the verification of how fast the myoelectric activity changes. The principle of the FFT algorithm is to decompose the sEMG signal into various sinus components of different magnitudes and frequencies. Another important measure in this study is the power spectral density (PSD) which is obtained by squaring the magnitude of each frequency component of FFT according to Equation 7.13 [21, 22].

$$PSD = |X(k)|^2 = \left|\sum_{n=0}^{N-1}sEMG(n)e^{-j\frac{2\pi kn}{N}}\right|^2 \tag{7.13}$$

and,

$$X(k) = \sum_{n=0}^{N-1}sEMG(n)e^{-j\frac{2\pi kn}{N}}, k = 0,1,..., N-1 \tag{7.14}$$

The relative changes in PSD of the sEMG signal during muscle voluntary contraction may be quantified by two well-known descriptors: Mean power frequency (MPF) and median frequency measure the central tendency and, then, indicate about which frequency the power of sEMG distributes. MPF and MDF indicators can be calculated by the following mathematical expressions, respectively [20] :

$$MPF = \frac{\sum_{f=0}^{\frac{f_s}{2}} f \times PSD(f)}{\sum_{f=0}^{\frac{f_s}{2}} PSD(f)} \tag{7.15}$$

$$\sum_{f=0}^{MDF}PSD(f) = \sum_{MDF}^{\frac{f_s}{2}}PSD(f) = \frac{1}{2}\sum_{f=0}^{\frac{f_s}{2}}PSD(f) \tag{7.16}$$

where f corresponds to the frequency bins represented in PSD and f_s is the sampling frequency.

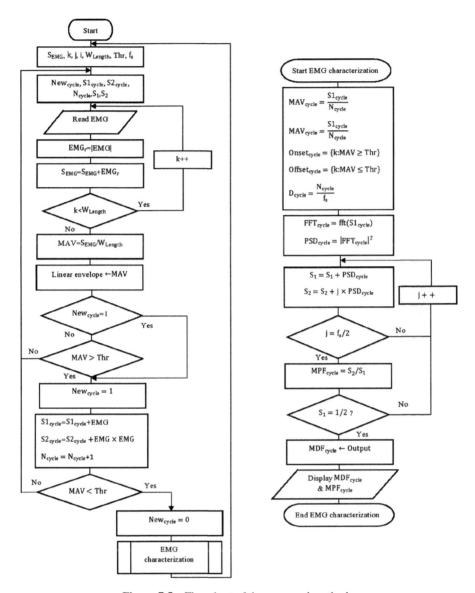

Figure 7.5 Flowchart of the proposed method.

7.4 Implementation Results

7.4.1 Implementation on compact RIO-9035 controller

The experimental setup in which the CompactRIO-9035 real-time controller and NI-ELVIS II+ board have been developed is depicted in Figure 7.6.

Figure 7.6   Experimental setup for hardware implementation.

CompactRIO is an industry-grade embedded controller manufactured by National Instruments. As an advantage, it has the capability of performing data acquisition and signal processing with a high level of precision. The NI-cRIO-9035 that has been selected for this study incorporates an in-built user reconfigurable FPGA (field-programmable gate array) type Xilinx Kintex-7 7K70T, eight slots for C Series I/O modules within one chassis for fast data acquisition and a real-time processor for high-speed data processing [23]. sEMG data are acquired through analog input module NI-9215 that has an external successive approximation register (SAR) 16-bit analog-to-digital converters (ADC). The host computer and NI-cRIO-9035 communicate via Ethernet port to deploy our application and to display the result from the NI-9474 digital output module. Technical characteristics of the cRIO-9035 and its used modules are enumerated in Table 7.1.

Figure 7.9 depicts a schematic diagram of the developed experimental setup. The NI-9215 is coupled to the FPGA integrated circuit. The FPGA can directly access the ADC acquired values and transfer them to the real-time processor for online processing, storage, and displaying the results. Within this procedure, two independent virtual instruments (vi) files are developed in LabVIEW software for the target FPGA and host real-time target. These .vi files are synchronized following the PCI (Peripheral Component Interconnect) bus interface. The role of the target FPGA .vi (Figure 7.7) is to acquire the sEMG signal with 5 sampling frequency. A further host real-time .vi is designed and run on a NI-cRIO- 9035 processor, as shown in Figure 7.8. The FPGA operates with a 40 MHz clock. Thus, the target FPGA .vi must be properly configured for executing such designated tasks. Configuration

Table 7.1   Technical characteristics of the NI-cRIO-9035 and its used modules.

Hardware name	Attribute	Specification
NI-cRIO-9035 [23]	CPU	Intel Atom E3825
	CPU frequency	1.33 GHz
	Number of cores	2
	RAM	4 GB
	Slot count	8-Slot
NI-9215 [24]	Module Type	Voltage Input
	Channels	4 Differential
	Sample Rate	
	Simultaneous	Yes
	Resolution	16-Bit
NI-9474 [25]	Module Type	Sourcing Output
	Channels	$8(DO_0, ...,DO_7)$
	Update Rate	1 μs

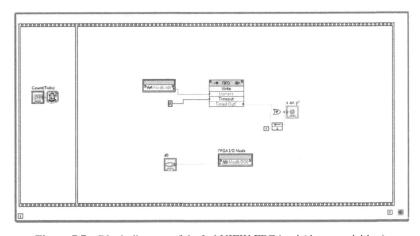

Figure 7.7   Block diagram of the LabVIEW FPGA .vi (data acquisition).

parameters used in FPGA .vi are presented in Table 7.2. The LabVIEW FPGA module enables Direct Memory Access (DMA) transfers of data to be performed with the use of FIFO architecture. The FIFO (First-In, First-Out) has consisted of two parts that act as one FIFO. The first one of this DMA-FIFO is designed on the FPGA using the FPGA's buffer. The second one of the DMA-FIFO is on the real-time controller. This part of the FIFO uses the host processor's memory resources. For every sample, the FPGA acquire value is put into a FIFO queue, which can be accessed from the real-time processor.

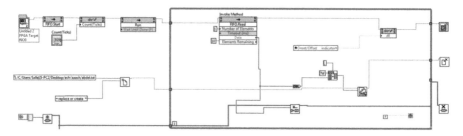

Figure 7.8   Block diagram of the LabVIEW host real-time.vi (data reading).

Table 7.2   Table 1. FPGA configuration parameters.

Attribute	Specification
Count(ticks)	8000
Name	FIFO
type	Target to host-DMA
FIFO's depths	1024
Data type	Fixed point
Word length	26 bits
Integer word length	5 bits

7.4.2 EMG instrumentation based on NI-ELVIS II+

NI-ELVIS II+ board is a hands-on design and prototyping platform that integrates the 12 most commonly used instruments including oscilloscope, digital multi-meter, function generator, bode analyzer, and more into a compact form factor ideal for the lab. This board has analog input channels of 8 differential or 16 single-ended ADC of 16-bit resolution and input frequency up to 50 to 60 Hz. NI-ELVIS II+ connects to a computer via USB and communicates using standard serial protocol [26]. After having built the EMG signal conditioning circuit on NI-ELVIS II+ as shown in Figure 7.10, the testing phase is required. For this purpose, we used the function generator of NI-ELVIS II+ to send EMG+ and EMG- signals shown in Figure 7.11, with a sampling rate of 5kHz to the input of the IA circuit (AD8226). The gain of the IA is fixed at 500 V/V via gain resistance of 100Ω. The output of the IA is fed into a high pass filter with a cut-off frequency of 20 Hz. The amplified sEMG signal from the final stage is input to the low-pass filter with a cut-off frequency of 2000 Hz.

7.4.3 Real-time evaluation

A computer application is designed for sEMG signal monitoring system based on LabVIEW software. LabVIEW is a graphical programming environment

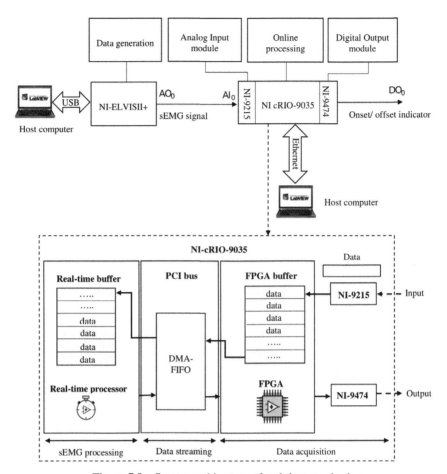

Figure 7.9   System architecture of real time monitoring.

that assists developers in easily building flexible and powerful test software with an intuitive GUI. In this study, the GUI is designed to illustrate and display the real-time results of sEMG analysis. Figure 7.12 describes the overall LabVIEW GUI of the study. After successful deploying of the host real-time .vi on the NI-cRIO-9035 processor, the users can validate the feasibility of the implemented system. A simple click on the 'Start acquisition & processing' button allows the users to monitor the variation of sEMG features in real-time during muscle contraction and relaxation movement. For more flexibility, the users have the privilege to start and stop the real-time acquisition data at any instant by clicking on the 'start/stop acquisition' button. In addition, configuration parameters related to data processing are available for the users to define customized sEMG control parameters. The LabVIEW GUI consists of

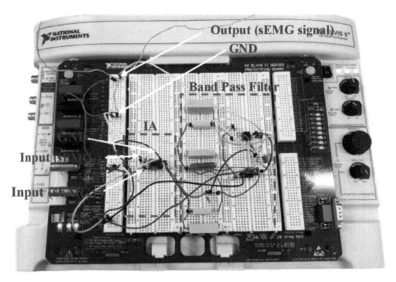

Figure 7.10   EMG signal conditioning circuit built on NI-ELVIS II + board.

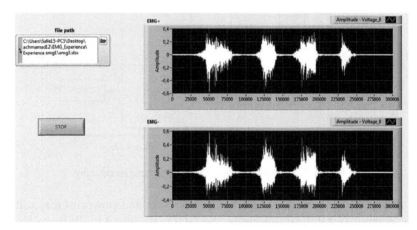

Figure 7.11   Raw sEMG signal (EMG+ and EMG−).

three different areas: (1) the parameters setting area, where the users can enter the count (ticks) to check data delivering frequency, and the window length together with the threshold value to avoid false alarms in onset/offset detection; (2) the data visualization area to display real-time monitored information in graphs (such as sEMG signal, linear envelope and PSD associated with the position of MDF as well as MPF); and (3) the computational display area for observing the key monitoring sEMG indicators (such as cycle duration, onset, offset, MAV, RMS, MDF, and MPF). Furthermore, the LED indicator

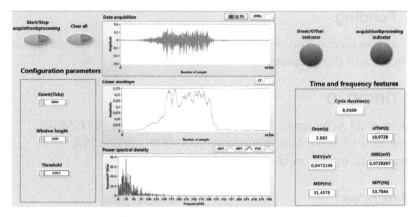

Figure 7.12   Front panel of the developed LabVIEW GUI.

lights up when the start of contraction occurs. Otherwise, the LED indicator turns off which indicates no occurrence onset event. This is also noticeable on the indicator of the NI-9474, as clearly shown in Figure 7.6.

The amplitude variations contained in the acquired sEMG signal are random, but it is certainly great medical information. In fact, it can inform the doctors about the state of the muscle. One of the main advantages of the sEMG signal is that it has temporal and frequency dynamics that are important to be quantified in order to characterize voluntary muscle contractions.

7.5 Conclusion

The goal of this chapter was to build an embedded system suitable enough for pre-processing, real-time acquisition, and processing of sEMG signals. The proposed investigation ended with the successful design and implementation of the EMG signal conditioning circuit on the NI-ELVIS II+ board, acquisition, and feature extraction through the CompactRIO-9035 controller. The physiological information extracted from each detected contraction/ relaxation cycle including time and frequency features were displayed in real-time on the LabVIEW GUI, providing a huge quantity of relevant health information for medical practitioners. Compared to commercial devices, our proposed system has proven to be highly adaptable to the needs of the medical community. The feasible outcome provided a novel technique to monitor sEMG information holistically, which can be potentially applied for further investigation on other surface EMG applications, especially medical diagnosis.

7.6 Funding

The authors received no financial support for the research, authorship, and/or publication of this chapter.

7.7 ORCID ID

Abdelouahad Achmamad: https://orcid.org/0000-0002-2951-5468s
Mohamed El Fezazi: https://orcid.org/0000-0001-6072-325X
Atman Jbari: https://orcid.org/0000-0002-1855-2503

References

[1] Chowdhury, Rubana H., Mamun BI Reaz, Mohd Alauddin bin Mohd Ali, Ashrif AA Bakar, Kalaivani Chellappan, and Tae G. Chang. Surface electromyography signal processing and classification techniques. Sensors. vol.13, no. 9, 2013, pp.12431–12466. https://doi.org/10.3390/s130912431

[2] Nazmi, Nurhazimah, Mohd Azizi Abdul Rahman, Shin-Ichiroh Yamamoto, Siti Anom Ahmad, Hairi Zamzuri, and Saiful Amri Mazlan. A review of classification techniques of EMG signals during isotonic and isometric contractions. Sensor. vol.16, no. 8, 2016, pp. 1304. https://doi.org/10.3390/s16081304

[3] Gerdle, Björn, et al. Acquisition, processing and analysis of the surface electromyogram. Modern techniques in neuroscience research. Springer, Berlin, Heidelberg, 1999, pp. 705–755. https://doi.org/10.1007/978-3-642-58552-4_26

[4] Rodríguez-Tapia, Bernabe, Israel Soto, Daniela M. Martínez, and Norma Candolfi Arballo. Myoelectric interfaces and related applications: current state of EMG signal processing–a systematic review. IEEE Access. 2020, pp. 7792–7805. https://doi.org/10.1109/ACCESS.2019.2963881

[5] Yochum, Maxime, Toufik Bakir, Romuald Lepers, and Stéphane Binczak. A real time electromyostimulator linked with emg analysis device. IRBM. vol.34, no. 1, 2013, pp. 43–47. https://doi.org/10.1016/j.irbm.2012.12.003

[6] Ma, Ruyi, Leilei Zhang, Gongfa Li, Du Jiang, Shuang Xu, and Disi Chen. Grasping force prediction based on sEMG signals. Alexandria Engineering Journal. vol. 59, no. 3, 2020, pp. 1135–1147. https://doi.org/10.1016/j.aej.2020.01.007

[7] Papagiannis, Georgios I., et al. Methodology of surface electromyography in gait analysis: review of the literature. Journal of medical engineering & technology. vol. 43, 2019, pp. 59–65. https://doi.org/10.1080/03091902.2019.1609610

[8] Jaramillo-Yánez, Andrés, Marco E. Benalcázar, and Elisa Mena-Maldonado. Real-time hand gesture recognition using surface electromyography and machine learning: A systematic literature review. Sensors. vol. 20, no. 9, 2020, pp. 2467. https://doi.org/10.3390/s20092467

[9] Jamal, Muhammad Zahak. "Signal acquisition using surface EMG and circuit design considerations for robotic prosthesis. Computational Intelligence in Electromyography Analysis-A Perspective on Current Applications and Future Challenges. no. 18, 2012, pp. 427–448. https://doi.org/10.5772/52556

[10] Ho, Tze-Yee, Yuan-Joan Chen, Wei-Chang Hung, Kuan-Wei Ho, and Mu-Song Chen. The design of EMG measurement system for arm strength training machine. Mathematical Problems in Engineering. 2015. https://doi.org/10.1155/2015/356028

[11] Mazzetta, Ivan, et al. Stand-alone wearable system for ubiquitous real-time monitoring of muscle activation potentials. Sensors. vol. 18, no.6, 2018, pp. 1748. https://doi.org/10.3390/s18061748

[12] Gentile, Paolo, Marco Pessione, Antonio Suppa, Alessandro Zampogna, and Fernanda Irrera. Embedded wearable integrating real-time processing of electromyography signals. In Multidisciplinary Digital Publishing Institute Proceedings. vol. 1, no. 4, 2017, pp. 600. https://doi.org/10.3390/proceedings1040600

[13] Chabchoub, Souhir, Sofienne Mansouri, and Ridha Ben Salah. Biomedical monitoring system using LabVIEW FPGA. World Congress on Information Technology and Computer Applications (WCITCA), 2015, pp. 1–5. https://doi.org/10.1109/WCITCA.2015.7367020

[14] Hsueh, Ya-Hsin, Chieh Yin, and Yan-Hong Chen. Hardware system for real-time EMG signal acquisition and separation processing during electrical stimulation. Journal of medical systems. vol. 39, no. 9, 2015, pp.1–8. https://doi.org/10.1007/s10916-015-0267-6

[15] Salman, Ali, Javaid Iqbal, Umer Izhar, Umar Shahbaz Khan, and Nasir Rashid. Optimized circuit for EMG signal processing. International Conference of Robotics and Artificial Intelligence. 2012, pp. 208–213. https://doi.org/10.1109/ICRAI.2012.6413390

[16] El Fezazi, Mohamed, Abdelouahad Achmamad, Mounaim Aqil, and Atman Jbari. PSoC-Based Embedded Instrumentation and Processing

of sEMG Signals. Analog Integrated Circuits and Signal Processing. 2021, pp.1–16. https://doi.org/10.1007/s10470-021-01850-x

[17] Data-sheets. Wide Supply Range, Rail-to-Rail Output Instrumentation Amplifier. https://www.analog.com/media/en/technical-documentation/data-sheets/ad8426.pdf

[18] Pasinetti, Simone, Matteo Lancini, Ileana Bodini, and Franco Docchio. A novel algorithm for EMG signal processing and muscle timing measurement. IEEE Transactions on Instrumentation and Measurement. vol. 64, no. 11, 2015, pp. 2995–3004. https://doi.org/10.1109/TIM.2015.2434097

[19] Merletti, Roberto, and Loredana R. Lo Conte. Surface EMG signal processing during isometric contractions. Journal of Electromyography and Kinesiology. vol. 7, no. 4, 1997, pp. 241–250. https://doi.org/10.1016/S1050-6411(97)00010-2

[20] Garcia, MA Cavalcanti, and T. M. M. Vieira. Surface electromyography: Why, when and how to use it. Revista andaluza de medicina del deporte. vol. 4, no. 1, 2011, pp. 17–28.

[21] Iscan, Zafer, Zümray Dokur, and Tamer Demiralp. "Classification of electroencephalogram signals with combined time and frequency features. Expert Systems with Applications. vol. 38, no. 8, 2011, pp. 10499–10505. https://doi.org/10.1016/j.eswa.2011.02.110

[22] Rechy-Ramirez, Ericka Janet, and Huosheng Hu. Bio-signal based control in assistive robots: a survey. Digital Communications and networks. vol. 1, no. 2, 2015, pp. 85–101. https://doi.org/10.1016/j.dcan.2015.02.004

[23] National Instruments. NI cRIO-9035 Manual. https://www.ni.com/fr-fr/support/model.crio-9035.html [accessed 20 January 2022].

[24] National Instruments. Ni 9215 Manual. https://www.ni.com/fr-fr/support/model.ni-9215.html [accessed 20 January 2022].

[25] National Instruments. Ni 9474 8 manual. https://www.ni.com/fr-fr/support/model.ni- 9474.html [accessed 20 January 2022].

[26] National Instruments.NI ELVIS II User Manual. https://www.ni.com/fr-fr/support/model.ni-elvisii. html. [accessed 20 January 2022].

SECTION 4

The Application of Embedded System in Image Processing

8

Quick and Efficient Hardware-Software Design Space Exploration Using Vivado-HLS: A Case Study of Adaptive Algorithm for Image Denoising

Sheetal U. Bhandari[1], Shruti S. Bansal[2], Pradip S. Thorbole[3], Rashmi P. Mahajan[4], and Priti J. Rajput[5]

[1]Dept. of E&TC, PCCOE, Savitribai Phule Pune, University, India
[2]Dept. of E&TC, PCCOE, Savitribai Phule Pune, University, India
[3]Sr. FPGA Design Engineer, Wipro Technologies, Pune, India
[4]Dept. of E&TC, MITAOE, Savitribai Phule Pune, University, India
[5]Dept. of E&TC, DYPSOEA, Savitribai Phule Pune, University, India
[1]sheetal.bhandari@pccoepune.org; [2]shruti9286bansal@gmail.com
[3]pradipthorbole@gmail.com; [4]dr.rashmimahajan@gmail.com;
[5]priti.rajput21@gmail.com

Abstract

The Field Programmable Gate Array (FPGA) devices provide ample configurable on-chip resources for the implementation of complex applications. The capability of these devices is enhanced by adding other blocks like multi-core processors, DSP blocks, on-chip block memories, etc. in basic FPGA fabric This makes it suitable for the implementation of real-time applications by using only Hardware (HW) or Hardware- Software (HW-SW) codesign approach. Advancements in FPGA demand upgrading implantation techniques in order to improve utilization capacity and the productivity of FPGA. One such technique is the ability to write applications in High-Level Languages (HLL) than the traditional Hardware Description Languages (HDL). It attracts HLL experts to the domain of FPGA without the need of studying complex HDLs.

The presented chapter explores Embedded System Design with Vivado HLS by implementing a case study on Least Mean Square (LMS) and

Normalized Least Mean Square (NLMS) algorithms for image denoising. The simplicity of coding provided by HLS has proved to be advantageous in quick turnaround of implementation without much knowledge of HDL. Available optimization directives with the tool have implemented the design with optimal resources and improved latency resulting in reduced computation cost.

8.1 Introduction

Field Programmable Gate Array (FPGA) devices have been an attractive option as accelerators for high computation demanding applications. However, many software engineers seldom choose FPGA for implementation due to two important factors, one is the requirement of hardware design knowledge and another is expertise in a programming language of FPGA. This imposes the need for a new approach for the programmers. Nowadays the programmers are originally trained to use High Level languages (HLL), which results in the demand for a technique that can convert HLL code to Register Transfer Level (RTL) for FPGA implementation. Smith et.al.[1] proposed a way to utilize the advantages of FPGAs without having the knowledge of its hardware by using high-level synthesis (HLS) tools.

HLS addresses this challenge; it accepts input as the algorithmic description of the application written in HLL like C/C++/SystemC etc. This algorithmic description is converted to the hardware description i.e., RTL netlist. An HLL description can typically be implemented faster and can reduce design efforts and susceptibility to programmer error. In recent years, HLS has enhanced its compatibility with input source code and produces more accurate output hardware designs. A recent study has proved that HLS allows the rapid evaluation of architecture regardless of the developer's expertise (software or hardware), thus expanding the horizons of the FPGA market [2]–[6]. There are various HLS tools available like Academic tools-LEGUP, DWARV, BAMBU, etc., and commercial tools like Altera SDK, Vivado HLS, etc. The comparison between academic tools and commercial tools is done by R. Nane et.al. [7] and the conclusion that was drawn based on the comparison was that commercial tools offer improved features and performance as compared to academic tools. In a comparison to Altera SDK and Vivado HLS, RTL developed by Vivado consumes less resources as compared to the RTL of Altera SDK.

Considering the market demands, the focus of this chapter is to explore Vivado HLS for the simplicity of coding and optimizations. Furthermore, this paper uses the case study of Least Mean Square (LMS) and Normalized Least Mean Square (NLMS) filter for image denoising [5]–[7]. For the implementation of the filters, the Zynq board is used and the results are presented in detail. Implementation is presented in three phases, wherein LMS and NLMS algorithms are designed in the first phase to proceed by its C code implementation in the second phase of the work. Finally, HW-SW with Zynq codesign aspect is explored in the third phase. The outline of this chapter is as follows: Section 2 deals with the basics of Vivado HLS, and the details of adaptive algorithm targeting LMS and NLMS filters are given in Section 3. Section 4 is devoted to the implementation of the LMS algorithm using Vivado HLS along with the discussion on results obtained during implementation [8]–[9]. Lastly, Section 5 provides the concluding remark and future scope.

8.2 High-level Synthesis

Designers can work at a higher level of abstraction with the help of HLS and software codes to specify the hardware. It is expected to provide high-performance and energy-efficient systems, in less time to market and can address today's system complexity. Traditionally, algorithm designers prefer to develop algorithms in high-level languages such as C/C++, and Vivado HLS is one of the tools capable of synthesizing C/C++ code into RTL for hardware implementation. However, many HLS tools are incapable of directly translating a high-level implementation to an RTL implementation [10]–[12]. In this case, to make the design synthesizable, developers must reorient the high-level implementations.

The HLS tool used in this work is Xilinx's Vivado HLS. Vivado HLS gives software engineers the flexibility and opportunity to accelerate designs with computational complexity on cutting-edge configurable devices [13]–[16].

C/C++ is one of the most popular programming languages used by software engineers to model algorithms. Even for hardware engineers with prior experience writing HDL, the HLS framework is a feasible alternative to hand-written HDL and is expected to help engineers improve productivity by enhancing the utilization of on-chip resources with less design time. As given in [3], the RTL developed by HLS is comparable to hand-coded RTL the biggest advantage of HLS RTL is its productivity. Any application developed in HLS gives six times higher productivity than hand-coded RTL. Figure 8.1 depicts how the Vivado HLS flow can replace HDL coding and behavioral simulation in the traditional FPGA design flow. Vivado HLS can be used to create RTL implementations by writing C/C++ code with proper restructurings and packing the design into Intellectual Property (IPs) for circuit integration. Vivado HLS has grown in popularity for modeling digital signal processing algorithms, including image processing algorithms. Furthermore,

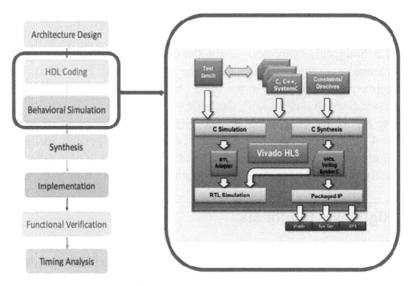

Figure 8.1   FPGA design flow with high-level synthesis.

Vivado HLS enables engineers to validate their designs for functional correctness using C/RTL co-simulation. Engineers will also be able to improve timing and performance by inserting optimization directives into the code and taking advantage of more parallelisms in the algorithm.

HLS offers many features such as,

1.   Reduced design efforts

2.   Easy verification

3.   Quick Design space exploration

4.   Easy adaption to new platforms

5.   Opportunity for software engineers to tackle hardware projects

Along with the above-mentioned features, it also reduces the design and verification time, furthermore reduces in development costs. As a result, the time to market is reduced and the use of hardware acceleration on heterogeneous systems becomes more appealing.

8.3 Adaptive Algorithm

In order to explore the possibilities of HLS, adaptive noise filtering algorithms are being selected as case studies. These filters will be used for de-noising

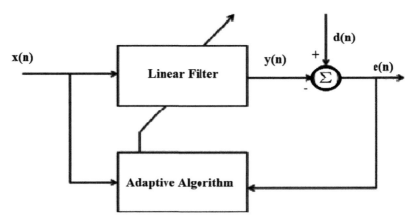

Figure 8.2   Block diagram of LMS filter.

real-time images collected from the camera. Any algorithm is chosen based on two key factors: its complexity and the number of computations. This section briefs about LMS and NLMS algorithms which are best-suggested candidates for such applications in literature [10]–[11], [17]–[23].

8.3.1 LMS algorithm

LMS algorithm is the most widely used algorithm for noise cancellation due to its simple architecture and less computation. The specialty of the LMS algorithm is that it requires less computational resources and memory. The study of the LMS algorithm is detailed in [17]–[23]. The basic operation of the LMS filter consists of the following two processes:

Filtering process - The output signal is produced based on the filter parameters and given input signal.

Adaptation and weight update process - The filter parameters adapt based on the environment of the error signal. The weights are updated for the FIR filter to minimize the error signal and bring the output signal close to the desired signal.

As shown in Figure 8.2, the LMS adaptive filter has a filtering section consisting of a linear FIR filter and an adaptation section consisting of the LMS algorithm.

Figure 8.2 shows that the noisy image is fed into the linear FIR filter. The output of the filter is compared to the desired image. The difference between the two is the error signal, which is used to update the filter's weights. Because the characteristics of noise are unknown in advance, adaptive LMS algorithms work best because they adjust their characteristics based on the current signals generated. The recursive LMS algorithm is so named because the current weight vectors are used to calculate the next iteration weight vector. The given formula is used to calculate the weight vectors for the next iteration.

$$\omega(n+1) = \omega(n) + \mu.x(n).e(n) \tag{8.1}$$

Where, x(n) = Input Signal, e(n) = error Signal, and μ = convergence factor

In the preceding equation, the choice of the convergence factor is critical in the design of a filter. The value of μ ranges from 0 to 1. The smaller the value of μ, the longer it will take to obtain the exact output. The larger the value of μ, the less time will take to get the bumped output. As a result, the value of μ should be determined by the application for which it is used. The value of μ in adaptive algorithms is chosen during the noise filtration process based on the value of the error signal. This allows the design to quickly adapt and produce noise-free output in a shorter amount of time.

8.3.2 NLMS algorithm

The Normalized Least Mean Square (NLMS) algorithm is a variant of the LMS algorithm. The major limitation of the LMS algorithm i.e., the fixed value of step size μ is tackled here. The input signal is never constant; hence the step size should be varied according to the varying input signal. This is said to be normalized step size and hence the Normalized Least Mean Square algorithm. The formula for step size calculation is given in 8.2.

$$\mu(n) = \frac{\beta}{c + |x(n)|^2} \tag{8.2}$$

Where,

μ = step size
β = normalized step size $(0 < \beta < 2)$
c = safety factor (small positive constant)

The formula for calculation of next iteration weight vector is given in 8.3,

$$\omega(n+1) = \omega(n) + \frac{\beta}{|x(n)|^2} x(n) \cdot e(n) \tag{8.3}$$

8.4 Implementation and Results

The implementation is carried out in 3 phases. In phase-1, LMS and NLMS algorithms are designed and the corresponding code in C is obtained from the Simulink model using Matlab Function Block. In Phase-2, C code of both models is imported, simulated, synthesized and optimized in Vivado-HLS. Based on a comparison of design metrics LMS algorithm is chosen for system implementation In Phase-3 a complete embedded system is developed around Zynq platform where real time images acquired by the Camera are de-noised using the LMS filter.

8.4.1 Phase I

The Simulink model of LMS and NLMS filter is given in Figure 8.3(a & b) respectively. The model consists of two images - the input noisy image and desired image. Since the LMS and NLMS function on 1D arrays, the 2D images are converted to 1D pixel values. The value of μ for the LMS algorithm is set to 0.01 as discussed in Section 3.1.

After successfully obtaining the output from the Simulink model, the C code of both the models is generated using the code generation function available in the configuration parameter tab in Simulink.

8.4.2 Phase II

In Phase-II, the C-codes from Simulink are imported and the RTL is generated using Vivado HLS. The C code is modified in two terms –

- From using "memset" instruction to using simple loops; as "memset" instruction was requiring more BRAM than available on board.

- Direct Memory Access (DMA) Access is modified to simple loops as DMA is not supported in HLS.

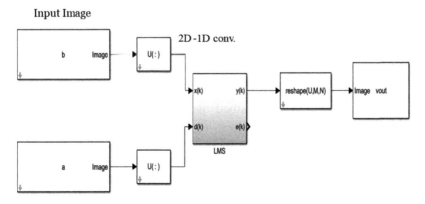

Figure 8.3(a)   Simulink model of LMS.

Input Image

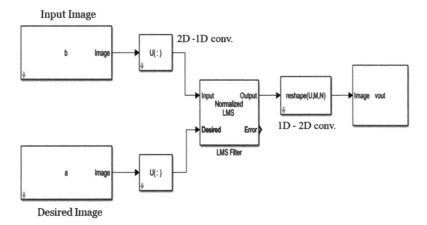

Desired Image

Figure 8.3(b) Simulink model of NLMS.

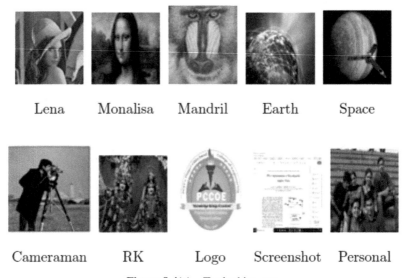

| Lena | Monalisa | Mandril | Earth | Space |

| Cameraman | RK | Logo | Screenshot | Personal |

Figure 8.4(a) Testbed images.

The RTL is successfully simulated. The simulation is carried out on 10 standard images. The testbed used for experimentation is given in Figure 8.4(a) & Table 8.1.

Design metrics for FPGA based systems are resource utilization and performance or latency. HDL experts use various modeling styles to ensure optimum fitting of functionality and the best possible performance by using HDL expertise. In order to extend this to HLL experts too, various optimization

Table **8.1** PSNR values of images.

Reference Image	PSNR Values		
	Input Image	Output Image	
		LMS	NLMS
Lena	13.5773	14.1637	14.2584
Monalisa	12.1408	13.3382	12.5043
Mandril	14.5607	15.4715	15.835
Earth	12.7931	14.5169	14.2563
Space	11.3661	12.9848	12.011
Cameraman	11.8246	14.2957	13.9192
RK	13.6197	16.5459	15.5612
Logo	12.9636	16.1005	13.7907
Screenshot	14.2325	20.206	15.8042
Personal	13.1032	14.4975	14.5587

Table **8.2** Summary of optimization directives.

Area of Improvement	Available HLS optimizations	Used in the design
Performance	Latency	Used
	Loop Flatten	Used
	Loop Merge	Used
	Unrolling	Used
Throughput	Dataflow	Used
	Pipelining	Used
Area	Scheduling	Use
	Binding	Used
	Inlining	Used
Handshake Protocols	Block level protocol	Used
	Port level protocol	Used

techniques are in-built in Vivado-HLS. Different optimization techniques available with Vivado and used for implementation are given in Table 8.2.

The resource utilization of LMS and NLMS filters with and without optimization is shown in Table 8.3. This table shows the impact of powerful optimization techniques available with the tool. It can also be observed that LMS needs less resources than the NLMS filter.

Another design metric is timing performance. Latency and initiation interval (Throughput) for both the algorithms are measured to select the best algorithm for final implementation. Table 1.4 shows that the latency and initiation interval of LMS is better than the NLMS algorithm.

In developing an embedded system on FPGA, the computational cost is calculated on two parameters, namely, resource utilization and latency as shown in Tables 8.3 & 8.4.

Table 8.3   Resource utilization.

Resources	Before Optimization LMS	Before Optimization NLMS	After Optimization LMS	After Optimization NLMS
BRAM_18K	7%	7%	7%	7%
DSP48E	43%	61%	43%	61%
FF	24%	27%	23%	26%
LUT	80%	92%	77%	89%

Table 8.4   Timing summary.

	Before Optimization				After Optimization			
Parameter →	Latency		Initiation Interval		Latency		Initiation Interval	
Algorithm ↓	Min	Max	Min	Max	Min	Max	Min	Max
LMS	35943598	35943598	35635202	35635202	10162316	421625616	10162302	421625602
NLMS	35943598	35943598	35635202	35635202	10444816	492134416	10444802	492134402

8.4.3 Phase III

Based on results obtained from Vivado HLS; the LMS algorithm is selected for system design as it provides better PSNR, Throughput over NLMS while taking fewer resources. The block diagram of the complete system is given in Figure 8.5.

The system is implemented on the Zybo board. Zynq platform is partitioned into ARM and FPGA fabric, i.e., Software implementation on ARM as a Processing system (PS) and hardware implementation on FPGA fabric as Programmable logic (PL). The LMS filter along with the other processing elements Simulink for HDMI camera are implemented as IP core onto the FPGA fabric. The other controlling signals are given by the PS via the AXI interface. This simultaneous working of PS and PL is referred to as Embedded System Design on FPGA.

The RTL of the system was developed with Vivado Design Suite 2018.2. wherein the RTL netlist one part is in the block of the LMS filter designed and imported from Vivado HLS. The system is simulated and synthesized successfully. The resource utilization of the system is shown in Table 8.5.

8.5 Conclusion and Future Scope

The domain of FPGA continues to advance in device heterogeneity with the ability to develop complete embedded system design on programmable chips (SoPC). Also, advanced design tools and techniques allow designers

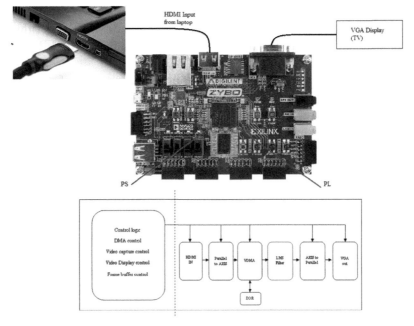

Figure 8.5 Block diagram of complete design.

Table 8.5 Resource utilization of the system.

Parameters	Slice LUTs (17600)	Slice Reg (35200)	Slice (4400)	LUT logic (17600)	DSPs (80)
Complete Design	9062	16109	4135	8512	7
Dynclk	219	341	104	219	0
Gpio btn	131	379	105	131	0
Gpio led	51	73	23	51	0
Gpio sw	132	379	87	132	0
Gpio video	93	31	33	93	0
Mem intercon	1809	3081	754	1549	0
Vdma	2586	4084	1063	2416	0
Dvi2rgb	421	472	165	397	0
Image filter	836	1071	338	836	7
Processing System	741	702	271	684	0
Rgb2vga	1	18	7	1	0

to develop systems on FPGA without expertise in HDL. For the software developers with expertise in of HLLs like C/C++ and less exposure to HDLs like VHDL/Verilog; implementation on FPGA turns out difficult. However, Vivado HLS provides the flexibility of coding in a high-level language to get RTL netlist generated automatically for ready deployment on FPGA. This leads to an improvement in productivity because of less design time to achieve a shorter time to market. The case study of LMS and NLMS algorithms for image denoising is successfully implemented on Xilinx Zynq FPGA using the HLS approach.

Other than the simplicity of coding and improvement of productivity, Vivado HLS also provides various "Optimization Directives".to improve the latency, performance, and resource utilization. After applying all the optimizations, resource utilization in terms of LUT was reduced by 3% and latency was improved from 35943598 to 10162316 for the LMS algorithm. It has been concluded that the implemented methodology reduces the computation cost of the application as it is calculated on resources and latency. Furthermore, it is observed that designing any application in Vivado HLS is less complex as compared to its counterpart VHDL/Verilog, as well as the flexibility of optimization can help a developer to optimize their design with respect to the four pillars of VLSI designs - area, speed, power, and performance.

To design and develop complex algorithms and complex IP development requires long-term design and verification efforts with great expertise. This research emphasizes the capabilities of HLS in complex algorithmic systems and emerging image processing fields. By using HLL in HLS, such complex algorithms can be easily developed and deployed on FPGAs. Although this research has not covered all capabilities of HLS. In the future, High-Level synthesis will play a major role to deliver ICs for fast-growing technologies where the time to market frame is very less. For examples, Video AI analytics, automatic driving cars, 5G communication, Advanced data centers, etc.

References

[1]   J. P. Smith et al., "A High-Throughput Oversampled Polyphase Filter Bank Using Vivado HLS and PYNQ on a RFSoC," in IEEE Open Journal of Circuits and Systems, vol. 2, pp. 241–252, 2021, doi: 10.1109/OJCAS.2020.3041208

[2]   J. Caba, F. Rincón, J. Barba, J. A. De La Torre, J. Dondo and J. C. López, "Towards Test-Driven Development for FPGA-Based Modules Across Abstraction Levels," in IEEE Access, vol. 9, pp. 31581–31594, 2021, doi: 10.1109/ACCESS.2021.3059941.

[3]   S. Lahti, P. Sjövall, J. Vanne and T. D. Hämäläinen, "Are We There Yet? A Study on the State of High-Level Synthesis," in IEEE Transactions on Computer-Aided Design of Integrated Circuits and Systems, vol. 38, no. 5, pp. 898–911, May 2019.

[4] I. Benacer, F. Boyer and Y. Savaria, "A High-Speed, Scalable, and Programmable Traffic Manager Architecture for Flow-Based Networking," in IEEE Access, vol. 7, pp. 2231–2243, 2019.

[5] T. Liang, J. Zhao, L. Feng, S. Sinha and W. Zhang, "Hi-DMM: High-Performance Dynamic Memory Management in High-Level Synthesis," in IEEE Transactions on Computer-Aided Design of Integrated Circuits and Systems, vol. 37, no. 11, pp. 2555–2566, Nov. 2018.

[6] Declan O'Loughlin et al., "Xilinx Vivado High Level Synthesis: Case studies" Research Gate, 2018.

[7] R. Nane et al., "A Survey and Evaluation of FPGA High-Level Synthesis Tools," in IEEE Transactions on Computer-Aided Design of Integrated Circuits and Systems, vol. 35, no. 10, pp. 1591–1604, Oct. 2016.

[8] Shaodong Qin, Mladen Berekovic, "A Comparison of High-Level Design Tools for SoC-FPGA on Disparity Map Calculation Example", 2nd International Workshop on FPGAs for Software Programmers, London, United Kingdom, September 1, 2015.

[9] Sharad Sinha, "Xilinx High-Level Synthesis Tool Speeds FPGA Design", Xcell Journal, Issue 83, 2013.

[10] Ajay Shiva, E. Senthilkumar, J. Manikandan and V.K. Agrawal, "FPGA Implementation of Reconfigurable Adaptive Filters", IEEE WiSPNET 2017.

[11] S. Haykin, "Adaptive Filter Theory", 3rd ed. Englewood Cliffs, NJ: Prentice Hall, 1996.

[12] Tom Feist , "White Chapter: Vivado Design Suite", June 2012.

[13] Ug871 and Ug902 for exploring vivado HLS.

[14] Ug937 for logic simulation.

[15] UG1270 (v2018.1), "Vivado HLS Optimization Methodology Guide" April 4, 2018.

[16] User Guide for "Introduction to High-Level Synthesis with VivadoHLS", 2016.

[17] Carlo Safarian, Tokunbo Ogunfunmi, Walter J. Kozacky, B.KMohanty, "FPGA Implementation of LMS-based FIR Adaptive Filter for Real Time Digital Signal Processing Applications", IEEE, 2015.

[18] Omid Sharifi Tehrani, Mohsen Ashourian, Payman Moallem, "FPGA Implementation of a Channel Noise Canceller for Image Transmission", IEEE, 2010.

[19] Shibalik Mohapatra, Asutosh Kar, Mahesh Chandra, "Advanced Adaptive Mechanisms for Active Noise Control: A Technical Comparison", International Conference on Medical Imaging, m-Health and Emerging Communication Systems, 2014.

[20] Gulden Eleyan, Mohammad Shukri Salman, Cemil Turan, "Two-Dimensional Sparse LMS for Image Denoising", 2015 Twelve International Conference on Electronics Computer and Computation (ICECCO), IEEE, 27–30 Sept. 2015.

[21] U. Meyer-Baese, "Digital Signal Processing with Field Programmable gate Array", 3rd edition, Springer.

[22] Atiq ur Rehman, Fahad Khan, Baber Khan Jadoon, "Analysis of Adaptive Filter and ICA for Noise Cancellation from a Video Frame", 2016 International Conference on Intelligent Systems Engineering (ICISE), IEEE, 15–17 Jan. 2016.

[23] D.B. Bhoyar; Soumita Bera; C.G. Dethe; M.M. Mushrif, "FPGA implementation of Adaptive filter for Noise Cancellation", 2014 International Conference on Electronics and Communication Systems (ICECS), IEEE, 13–14 Feb. 2014.

9

Fast FPGA Implementation of A Moving Object Detection System

**Mohamed Sejai[1], Anass Mansouri and
Saad Bennani Dosse[2], Yassine Ruichek[3]**

[1]Intelligent Systems, Georesources and Renewable Energies Laboratory,
Sidi Mohamed Ben Abdellah University, Faculty of Sciences and Technics
Fez, Morocco.
[2]Sidi Mohamed Ben Abdellah University, National School of Applied
Sciences Fez
[3]Belfort-Montbéliard University of Technology, France
[1]mohamedsejai92@gmail.com; [2]Morocco.anas_mansouri@yahoo.fr and
bennani.saad.ensaf@gmail.com; [3]yassine.ruichek@utbm.fr

Abstract

Real-time object detection has become a popular technology in many applications such as real-time security monitoring, robot navigation, event detection, and autonomous vehicles. In this work, we propose a real time implementation method of detecting moving objects from a video stream, the main element of the used method is based on the comparison of different video frames. The complete video motion detection chain has been implemented in the software and hardware by using monocular static vision which requires high computational performance, the hardware implementation of the proposed system reduces the detection time of moving objects compared to the software implementation and achieves a processing speed of 30 fps for a resolution of 640 × 480 pixels. Finally, several tests are performed in real time and the experimental results show that the proposed method is robust and provides excellent performance in terms of detection accuracy and processing speed, which is suitable for real time applications.

9.1 Introduction

Detecting moving objects from video footage becomes necessary to make analysis more intelligent for a number of embedded applications such as security, vehicle navigation, and traffic control, etc. The detection of a moving object has several constraints related to the processing time of different objects on the one hand and the performance of the detection on the other hand which depends on many parameters such as light variation, presence of shadows, and others [1].

The main objective of moving object detection is to distinguish the foreground of the object from the stationary background. Several methods have been proposed so far for moving object detection, including background subtraction, image differentiation, temporal differentiation, and optical flow [2, 3, 4, 5]. The most common approach for detecting moving objects is based on the image subtraction technique. When a new image is captured, the difference between the image and the next image is computed to extract the difference image that marks the areas where a moving object was in the image, this subtraction calculation of pixel intensity is simple and easy to implement [6].

This work is based on image differences for moving object detection and tracking using a static camera. The processing chain contains RGB to grayscale conversion block, subtraction between images, thresholding, noise removal filter, and edge detector of moving objects in the scene. This detection algorithm of moving objects is greedy in calculation and carries out treatment with the delayed time, the purpose of this work is to reduce the time of treatment of the moving objects detection process based on a technology of Hardware implementation on FPGA.

For example, several hardware implementations have been proposed in the literature to improve the quality of the image and accelerate the processing time for the requirement of the critical system. Siva Nagi Reddy Kalli and Bhanu Murthy Bhaskara [7] implemented a real-time Moving Object Segmentation in FPGA by using Background Modeling with Biased Illumination Field Fuzzy C-Means, the system response produced accurate results. In the paper [8], the authors proposed a Visual Object Tracking by Adaptive Background Subtraction on FPGA to obtain a more complete moving object, in this work, the threshold gives better noise immunity, and to eliminate the noise a morphological filter is used. I.Iszaidy et al [9] proposed an analysis of background subtraction on embedded platform based on synthetic dataset. The proposed system is to provide a comparative analysis of available background subtraction algorithms on the embedded platform. In the work referenced in [10], the authors presented the FPGA based Object

Detection using Background Subtraction and Variable Threshold Technique and demonstrated the ability of object detection using background subtraction with the variable threshold method.

In this chapter, our used algorithm is implemented and its performance is measured. This algorithm is presented in the next section. We first propose a software implementation of our algorithm based on the image subtraction method on the Raspberry Pi4 platform, and then we propose hardware implementation based on a reconfigurable FPGA circuit, the Altera Cyclone IV FPGA, in order to accelerate the processing time and to meet the real-time processing constraints. The key element of this work is based on the proposed the fast architecture of the moving objects detection. A performance comparison between the two types of implementation is performed to prove their efficiency. The used system in this chapter succeeded in detecting moving objects accurately in video sequences, it meets the requirements of real-time and accuracy.

The remainder of this chapter is organized as follows. Section 2 gives an overview of the Detecting Moving Objects algorithm and detailed explanations for each stage. The hardware implementation of the system is represented by its main module blocks and the experiment result from the presented moving object detection algorithm is presented in section 3. Finally, Section 4 concludes the chapter and discusses future works.

9.2 Detect Moving Objects Algorithm

The algorithm used in this work is based on a vision system that can capture the video and process it in order to have in the last step of the processing chain a final output in the form of a video for the detection and tracking of moving objects. With a fixed environment, the background image is selected as the black image, the following images contain the updated location of the target.

The image differentiation method identifies the presence of a moving object by considering the difference between two consecutive images, i.e., the changes between consecutive images caused by a moving object are taken into account to distinguish the foreground from the background in the scene. The organization of the used system is as follows:

1. To extract the moving object from the video image and obtain its characteristic information, the Subtraction between images is applied.

2. In order to accurately reflect moving objects in the image, the threshold is used to improve the image for meeting the needs of high-quality moving objects detection.

3.   The performance of the image is achieved by a median filter in order to reduce the noise and to eliminate everything in the resulting image which is not the part of interest.

4.   The edge detection operator is used to detect the edge of the moving object, which greatly improves the segmentation accuracy.

The overall process of the moving object detection algorithm can be explained in a schematic way as shown in Figure 9.1 to detect the moving objects in a video sequence.

Figure 9.1   Processing chain of the algorithm used to detect moving objects based on the difference between images.

We convert all the images in the color video sequences to gray scale images in order to have better performance and to reduce the computational requirements. This is because the color image increases the amount of data processing to achieve good performance because a pixel in RGB format is represented by three 8-bit data, however, for a 640 × 480 pixel image in RGB format contains 921600 information bytes. In the literature, many approaches for color to gray scale conversion have been proposed [11]. A common approach for the conversion of RGB to gray scale is expressed by the formula (9.1).

$$I = 0.299 * R + 0.587 * G + 0.114 * B \qquad (9.1)$$

The technique of frame subtraction [12], detects and tracks the object over time by locating its position by subtracting consecutive frames from the video, moving objects, which are called foreground, from the background of the sequence in the video stream are detected without having any prior information about the object. We consider $I_i(x,y)$ that is the intensity of the I_i frame. The difference image $I_d(x,y)$ used to detect the moving region is calculated mathematically by the expression (9.2):

$$I_d(x,y) = |I_{i+1}(x,y) - I_i(x,y)| \qquad (9.2)$$

where x and y are the pixel coordinates.

In order to meet the needs of high quality moving objects in various fields, the threshold is used to segment the image, if the pixel value is higher than or equal to the threshold value (T), the pixel is considered to belong to the background. Otherwise, it is considered to belong to a moving object, so that the binary image can accurately reflect the moving objects of the image in which the pixel of 0 is determined as the background region, and the pixel of 1 is determined as the moving target region [13]. A classification pixel into object pixel or background pixel is based on the following decision:

$$I_T(x, y) = \{0: I_d(x, y) \leq T; 1: \text{otherwise}\} \tag{9.3}$$

Where $I_T(x,y)$ is the binary image at the coordinates x and y.

The threshold image obtained contains the motion region along with noise. Therefore, noise needs to be removed [14]. In this chapter, median filter is applied to 3×3 Window size for smoothness and for removal of noise and eliminates pixels that are not related to moving objects.

Image segmentation is an important method in computer moving object detection and tracking. The edge detection operator is applied to find the edge of the detected moving objects, which greatly improves the segmentation accuracy. The sobel method uses two kernels measuring 3×3 pixels to calculate the gradient [15]. The gradient calculation is based on the following expression:

$$G = \sqrt{G_X^2 + G_Y^2} \tag{9.4}$$

Where G: is the Sobel gradient operator value, G_X: Horizontal sobel gradient and G_Y: Vertical sobel gradient.

The proposed system can experiment with different metrics which can be used for performance evaluation.

9.3 Implementation and Experiment Results

The algorithm proposed in this work is implemented in two different ways to verify the accuracy of the system. In the first experiment, a software implementation of the proposed technique is performed on a Raspberry Pi4 platform. In the second experiment, the proposed architecture is implemented in a hardware manner on an FPGA board. In both experiments, we tested the performance of our method using a set of 640×480 video sequences to verify the efficiency of our proposed algorithm. The proposed system detects moving objects and tracks them correctly and continuously.

9.3.1 Software simulation and evaluation

To evaluate the performance of our moving objects detection algorithm, a software simulation was implemented on a Raspberry Pi4 under Linux operating system and the software implementation was developed in C programming language and the Open Computer Vision (OpenCV) 4.x library to verify the system output.

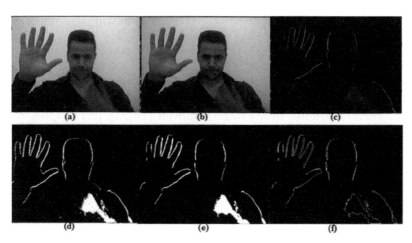

Figure 9.2   (a) Video to be tested (b) gray scale video version (c) frame Subtraction result (d) thresholding video (e) median filtered video (f) edge detection to obtain the result of the proposed method.

Figure 9.3   (a) Video to be tested (b) gray scale video version (c) frame Subtraction result (d) thresholding video (e) median filtered video (f) edge detection to obtain the result of the proposed method.

As presented above, the moving object detection system provides information about the moving objects in the scene. Figures 9.2 and 9.3 show the different objects extracted from the video stream. The inputs of the algorithm are the frames of the sequences. The obtained results show that it is perfectly possible to detect objects in the scene from the cameras.

The extraction of moving objects from the video is based on the frame subtraction method. Despite several drawbacks of this method, such as camera noise, shadows, light, etc., this algorithm can detect the moving object in an effective way. Some pixels of the object become part of the background, and the quality of the foreground object is highly dependent on the threshold value. The method used in this work can detect moving objects by removing the noise. By analyzing the obtained results, we conclude that the algorithm works well with an improvement of the results by using an image processing filter and an edge detector.

The processing time of the overall detection chain time is 3.8s/frame. The program takes longer to execute due to its iterative structure and the number of memory allocations it performs. The execution time on the Raspberry Pi 4 platform is more important due to the constraints of the platform used. The main disadvantage of this software solution is low performance in terms of processing speed. As a result, detecting moving object consumes very high computation time, making this solution unsuitable for applications that require real-time processing. To overcome this problem, the solution requires specially designed hardware implementation to perform the task in a high-speed processing based on a reconfigurable FPGA circuit.

9.3.2 Embedded objects detection system

The used algorithm for moving object detection has been developed and implemented on an FPGA board, which provides Altera Cyclone IV chips with the aim of improving the processing time of a moving object. The block diagram for object detection and tracking of the proposed implementation is shown in Figure 9.4. Additionally, a display device is used to show the system outputs. The processed data is displayed on the VGA screen. The design, simulation, and implementation were developed using Quartus. The system consists of the hardware design of the algorithm by using the VHDL as a hardware description language.

The proposed system of the moving object detection algorithm is illustrated in Figure 9.4, which presents the hardware design of the system. The hardware implementation is composed of the video acquisition, blocks for moving object detection, and the display of the processed data on the VGA output. In the acquisition and pre-processing stage, the decoder board is configured to digitalize the incoming composite video, its configuration is given through I2C protocol communication, One of the advantages of using the

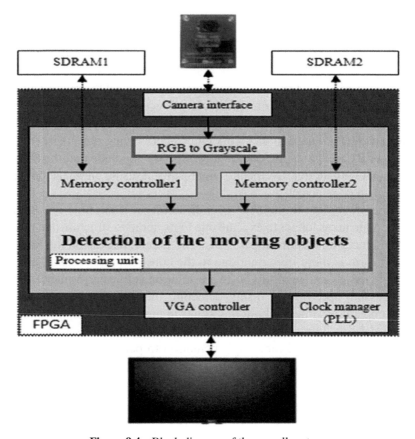

Figure 9.4 Block diagram of the overall system.

camera is the possibility of changing its configurations, such as image resolution, frame rate, etc. The digitalized video is converted to RGB format. The motion of the object is detected and localized by processing the video with our image subtraction algorithm, after this process we defined each pixel, either as black: non object pixel, white: object pixel. Finally, the processed video of the moving object is displayed on a VGA screen.

To meet a real time implementation of a moving objects detection system, the hardware system of the used algorithm is proposed to have good performance. In the proposed system, the relations between the principal blocks are: The both RGB images, image i and image i+1 are first converted into gray scale format and then stored in the two external SDRAM memory of the FPGA platform that is used as a video buffer to apply the frame subtraction method to detect the moving object and then to black and white the video sequence by the developed hardware threshold block. A block of image processing through a fast median filter module for reducing the undesired effects caused by the illumination is also developed, finally a sobel edge detector is used to find the edge of the detected moving objects, and the displayed image will show the white moving object. The obtained results are quite satisfactory, as shown in Figure 9.5.

Figure 9.5 (a-f) depict the obtained results of the different hardware implementations which is the process of detecting moving objects in the static scene from a video sequence captured by a static camera. For verifying the output data of our system, various videos are tested. There's some noise in the image color (a) but quite reduced in the image color (c) by using the median filter. The image in gray scale format is shown in the image (b). Figures (d) and (e) illustrate respectively the results of the thresholding image and the sobel edge detector. As you can see from the figure (f), the output of the proposed system was successfully detecting the moving objects with a minimal amount of noise.

According to our performance evaluation results, the hardware implementation confirms that the proposed hardware system process approximately 30 frames per second (fps) on average and accurately detects the moving objects, and have similar to the software implementation which processes

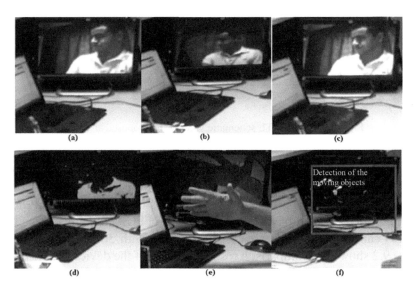

Figure 9.5 (a) RGB video (b) RGB to gray scale conversion (c) median filter result (d) thresholding video (e) edge detection (f) result of the proposed method.

approximately 1 frame per 3.8s. The second implementation is faster than the first one; however, it provides an image of a detected moving object in real time.

Table 9.1   SW and HW results for VGA resolution.

	Software	**Hardware (fps)**	**speed**
Detect moving object system	1 frame / 3.8 s	30 frames / 1 s	114

The implemented algorithm on the FPGA has significant improvement in the object detection process due to the parallel operations that effectively reduced the processing time. We conclude that the FPGA implementation improves the video processing algorithm 114 faster than the raspberry pi4 implementation as shown in Table 9.1, which satisfies the requirement of real-time processing.

The proposed hardware implementation is designed using the VHDL language for coding. Our design was synthesized for the FPGA cyclone IV as shown by the RTL schematic view in Figure 9.6.

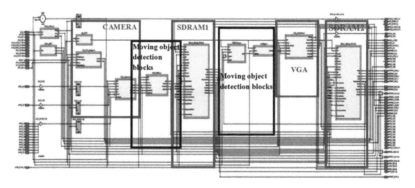

Figure 9.6   The RTL schematic view of the proposed architecture.

The proposed moving object detection technique gives a good result in terms of image quality and processing speed compared to similar works in recent year. In addition, the parallel operations effectively reduced the processing time of the proposed architecture. The FPGA implementation of the system can detect moving objects in real time with 30 frames per second. Table 9.2 shows the comparison results in terms of the device, frame per second, algorithm, and frame size used by similar works.

Table 9.2   Comparison with other works.

Reference	Device	FPS	Algorithm	Frame Size
[16]	Virtex 2 Pro FPGA	30	background subtraction	720*480
[17]	Altera Cyclone IV FPGA	30	Frame differencing	640*480
[18]	Xilinx Virtex-5 FPGA	27.2	Optical Flow	480 × 704
[19]	Xilinx Zynq FPGA	25	background subtraction	–
Our work	Altera Cyclone IV FPGA	30	Frame subtraction	640*480

9.4 Conclusion

In this chapter moving objects are detected based on the frame subtraction method. This chapter is focused on the video which is captured from a static camera. Two experiments for moving objection detection are described. The first is implemented on Raspberry pi4 by using c programming language with Open CV Library, while the second is implemented on FPGA, the proposed architecture is realized using VHDL, The results of the prototype system are obtained at 640×480 resolution, the proposed architecture can be employed in various real-time computer vision applications and FPGA-based system is a good solution in real-time computer vision problem. Finally, the results of the multiple videos test show that the proposed technique performs well, eliminate noise and detect moving object more completely for both indoor and outdoor sequences. In future, this work can be extended to a moving camera.

References

[1]   Jaya S. Kulchandani, Kruti J. Dangarwala. « Moving Object Detection: Review of Recent Research Trends". International Conference on Pervasive Computing, 2015.

[2]   Soharab Hossain Shaikh, Khalid Saeed , Nabendu Chaki."Moving Object Detection Using Background Subtraction". 2014.

[3]   Yong, H.; Meng, D.; Zuo, W.; Zhang, K. Robust online matrix factorization for dynamic background subtraction. IEEE Trans. Pattern Anal. Mach. Intell. 2018.

[4]   S. S. Sengar, S. Mukhopadhyay, "Moving object detection based on frame difference and W4" in Signal, Image and Video Processing 2017; 11 (7): 1357–1364.

[5] X. Han, Y. Gao, Z. Lu, Z. Zhang, D. Niu, "Research on moving object detection algorithm based on improved three frame difference method and optical flow" in Fifth International Conference on Instrumentation and Measurement, Computer, Communication and Control (IMCCC); Qinhuangdao, China; 2015. pp. 580–584.

[6] N. Singla, "Motion detection based on frame difference method" in International Journal of Information and Computation Technology 2014; 4(15): 1559–1565.

[7] Siva Nagi Reddy Kalli and Bhanu Murthy Bhaskara "Implementation of Moving Object Segmentation using Background Modeling with Biased Illumination Field Fuzzy C-Means on Hardware Accelerators". Asian Journal of Scientific Research, 2017.

[8] Denna Paul and Ranjini Surendran, "Implementation of Visual Object Tracking by Adaptive Background Subtraction on FPGA", International Journal of Engineering Research & Technology, 2015.

[9] I.Iszaidy et al, "An Analysis of Background Subtraction on Embedded Platform Based on Synthetic Dataset", 5th International Conference on Electronic Design, 2020.

[10] Sridevi N and M Meenakshi2, "FPGA based Object Detection using Background Subtraction and Variable Threshold Technique", International Journal of Applied Engineering Research, 2018.

[11] Ali Günes, Habil Kalkan and Efkan Durmu, "Optimizing the color-to-grayscale conversion for image classification", Signal Image and Video Processing, 2016.

[12] Clement Alabi and James Ben Hayfron-Acquah "An Improved Frame Difference Background Subtraction Technique for Enhancing Road Safety at Night", International Journal of Computer Applications, 2021

[13] Manikanta Prahlad Manda and Hi Seok Kim "A Fast Image Thresholding Algorithm for Infrared Images Based on Histogram Approximation and Circuit Theory". MDPI, 2020.

[14] Angalaparameswari Rajasekaran and Senthilkumar Palaniappan, "Image Denoising Using Median Filter with Edge Detection Using Canny Operator", International Journal of Science and Research, 2014.

[15] Darma Setiawan Putra et al, "Implementation of Sobel Method Based Edge Detection for Flower Image Segmentation", Journal Publications & Informatics Engineering Research, 2019.

[16] A. Lopez-Bravo et all, "FPGA-based video system for real time moving object detection". International Conference on Electronics, Communications and Computing. 2014

[17] Jia Wei Tang, Nasir Shaikh-Husin, Usman Ullah Sheikh, and M. N. Marsono, "FPGA-Based Real-Time Moving Target Detection System for Unmanned Aerial Vehicle Application". International Journal of Reconfigurable Computing. 2016.

[18] Jaechan Cho , Yongchul Jung , Dong-Sun Kim , Seongjoo Lee and Yunho Jung , "Moving Object Detection Based on Optical Flow Estimation and a Gaussian Mixture Model for Advanced Driver Assistance Systems". MDPI. 2019

[19] Masayuki SHIMODA , Shimpei SATO, and Hiroki NAKAHARA, "Power Efficient Object Detector with an Event-Driven Camera for Moving Object Surveillance on an FPGA". 2019

10

Face Recognition based on CNN, Hog and Haar Cascade Methods using Raspberry Pi v4 Model B

H. Belmajdoub, R. Mafamane, Y. Bekali Karfa, M. Ouadou and K. Minaoui

LRIT Associated to the CNRST-URAC n°29, Mohammed V University, Rabat, Morocco.
Hanae_belmajdoub@um5.ac.ma, rachid_mafamane@um5.ac.ma, y.bekali@um5r.ac.ma.com, mr.ouadou@um5r.ac.ma, khalid.minaoui@fsr.um5.ac.ma

Abstract

Automatic person recognition has received a lot of attention in the last few years. It is one of the best biometric modalities for applications related to the identification or authentication of persons. Since face recognition is a compute-intensive process, an embedded system solution allows for low-cost, discrete hardware implementations that can be applied and embedded in a wide range of applications. Several methods have been developed for face detection such as CNN, Haar Cascade, and Hog. In this work, we will apply an algorithm that exploits its different methods for facial recognition based on a powerful embedded system and connected object such as the Raspberry Pi. This study aims to better understand the difference between each method and to compare the effectiveness of each learning technique used.

10.1 Introduction

In recent times and with the advancement of technology, passwords and keys used in all areas of security and access control have become easily forged and breached. This is how biometrics was invented and became fashionable in areas that require a high level of security and control. Among all the

195

biometric technologies that exist, facial recognition is one of the most used and adapted. It allows exploiting a lot of information about a person.

Facial recognition is applied mainly in the field of security [12]. It is responsible for the identification and authentication of the face [8]. It can also be used to detect, track and recognize persons in public areas like shopping payment, schools, ministries, airports, areas with restricted access such as private offices, houses... etc

Facial recognition [14] is a technology based on biometric techniques, artificial intelligence, 3D mapping, and deep learning. This technology allows identifying a person on an image or video frame, by comparing the characteristics of his face with those saved in the dataset automatically.

The facial recognition system used in this chapter is based on the Raspberry Pi 4 Model B. It is a small board, which can be connected to additional modules. It is a good platform for testing the functionality of a connected object before it is put into use [11].

Face recognition is sensible to consider an embedded system implementation that would be specifically enhanced to recognize faces. An embedded system would provide multiple benefits, such as low cost, since just a small subset of hardware is required relative to general-purpose computing solutions, improving face recognition processing algorithms in real time independently of the other post processing issues, and integrating with other technologies.

The Raspberry Pi is used to be connected and controlled by external devices and aims to create Internet of Things (IoT) solutions.

The IoT [10] is a system of interconnection between computer devices, machines, objects, animals, and even people, provided with unique identifiers (UID) with the ability to transfer data over a network. Technically, this interconnectivity digitally identifies an object through a wireless communication system, such as wi-fi or bluetooth. And without human-to-human or human-to-computer interaction. Collectively, it is any natural or man-made "object" to which an IP address can be assigned and which can transfer data over a network (Figure 10.1).

The domain of facial recognition has suffered many challenges since its development, with various reasons making accurate recognition a difficult task, as shown in the following:

- Illumination: Changing lighting levels can affect the image of a person's face in a variety of ways.

- Occlusion: Once a part of the face is hidden, the facial characteristics cannot be totally visualized, and as a result authentication by the facial recognition application is compromised.

Figure 10.1 Raspberry Pi board.

- Expressions: Face recognition is also affected by facial expressions, resulting from the motion of human facial characters, causing modifications to the facial picture. Therefore, certain face recognition applications cannot recognize the various expressions of an identical person, which implies that a problem may occur in the recognition of a person's face.

- Hair: Human hair frequently overlaps the forehead. Therefore, in the majority of face recognition solutions, hair is removed from the pictures in the dataset to avoid it acting as a face recognition obstacle.

Several studies have used facial recognition to identify a person using the Raspberry Pi. Authors in [9] have proposed a face recognition system based on the Raspberry Pi board. The system detects faces using Haar feature-based cascade classifier. In fact, the authors in [7] used conventional face detection and recognition techniques such as Haar and PCA detection. The chapter [13] proposed to use Raspberry Pi, cloud, and a wireless camera for face recognition. The method uses a bag of words for the extraction of oriented FAST points and rotated BRIEF points from the detected face. Then it uses a support vector machine for identification. Another chapter [16] deployed a group of computers connected to a microcomputer with a camera to detect and recognize a face in different situations. This system uses "Boosted Cascade of Simple Features" for face detection, and "Local Binary Pattern algorithm" for recognition. Authors in [5] applied the Eigenface vector algorithm for face recognition. It is characterized by less computation and fast execution, which makes it ideal for Raspberry Pi. Chapter [3] created a detection system using an RPi integrated intelligence with an HD camera to process images and detect motion. On the other hand, authors in [4] have used a convolutional neural network (CNN) for face recognition from a set of images tracked in a video. Authors in [6] have proposed gender detection based on computer vision and machine learning using a convolutional neural network (CNN) which is used to extract various facial features. The realization is done on a Raspberry Pi board programmed using Python.

In the present work, we implement a face recognition application based on the Raspberry Pi4 single board computer. The facial recognition system includes face detection and localization is done using different algorithms such as Haar Cascade, Hog, and CNN.

The goal of this application is to know the difference between each method used for facial recognition. At the end of the project, we will compare the results obtained from the different algorithms used.

This chapter is organized as follows: Section II presents the different methods used, while Section III describes the application and reports the results and analysis. Finally, a conclusion is drawn in Section IV.

10.2 Implementation Methods

The objective of this chapter is to deploy the different learning methods for facial recognition using a Raspberry Pi 4 model B+ board with Pi2 camera module. These methods are Haar Cascade, Hog and CNN.

10.2.1 Method 1. Haar cascade

Haar Cascade classifiers are an effective way for object detection. This method was proposed by [15]. It is a machine learning based approach where a cascade function is trained from a lot of positive and negative images. The positive images contain the images which we want the classifier to identify. The negative Images are of everything else, which do not contain the object to be detected. A cascade classifier makes large numbers of small decisions whether it's the object or not. The structure of the cascade classifier is represented in (Figure 10.2).

10.2.2 Method 2. Histogram of Oriented Gradients (HOG)

HOG is a feature descriptor generally used for object detection. HOGs are widely known for their use in pedestrian detection. A HOG relies on the

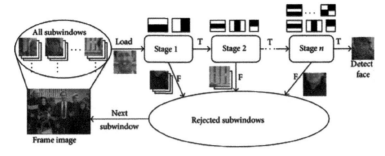

Figure 10.2 Architecture of haar cascade.

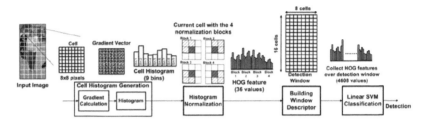

Figure 10.3 Architecture of hog.

vproperty of objects within an image to possess the distribution of intensity gradients or edge directions. Gradients are calculated within an image per block. A block is considered as a pixel grid in which gradients are constituted from the magnitude and direction of change in the intensities of the pixel within the block [1] (Figure 10.3)

10.2.3 Method 3. Convolutional Neural Networks (CNN)

CNN or ConvNet, [2] are a variant of artificial neural networks whose layer types and organization differ from multilayer perceptron's. They are also part of supervised learning methods, capable of performing, in addition to classification, the extraction of characteristics from images. They are used in the field of imaging since their appearance. Their architecture allows the solution of both problems: On the one hand, they keep the two-dimensional shape of the image, which is essential to extract its details and characteristics; on the other hand, they use small matrices called "Filters" with the convolution operation to browse and make changes to the image. It is less expensive in terms of computation and RAM storage compared to artificial neural networks (see Figure 10.4).

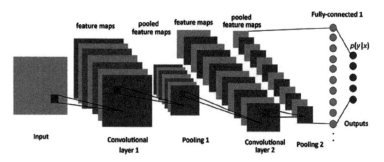

Figure 10.4 Architecture of CNN.

10.2.3.1 Convolution layer

The convolution layer (CONV) uses filters that scan the input according to its dimensions by performing convolution operations. It can be set by adjusting the filter size and the stride. The output of this operation is called a feature map or activation map.

10.2.3.2 Pooling layer

The pooling layer (POOL) is a subsampling operation, typically applied after a convolutional layer. In particular, the most popular types of pooling are max and average pooling, where the maximum and average values are taken, respectively.

10.2.3.3 Fully connected layer

The fully connected layer (FC) is applied on a previously flattened input, where each input is connected to all neurons. Fully connected layers are typically present at the end of CNN architectures and can be used to optimize objectives such as class scores.

10.3 Deployment Environments and Results

The objective of our application is to realize a connected object for facial detection and recognition using the Raspberry Pi board based on different learning models.

The hardware and software platform used to realize our application and the structure of the algorithm are as the following.

10.3.1 Hardware environment

10.3.1.1 Raspberry Pi4

With 4 GB of memory, this Raspberry PI 4 is the most powerful PI 4 yet retains exceptional capabilities compared to the Pi 3 and Pi 3 Model B+.

It offers an unprecedented increase in processor speed, multimedia performance, memory, and connectivity over the previous generation Raspberry Pi 3 Model B+ while maintaining compatibility with previous versions and similar power consumption. For the end-user, the Raspberry Pi 4 Model B offers desktop performance comparable to entry-level x86 PC systems.

Key features of this board include a high-performance 64-bit quad-core processor, dual display support at resolutions up to 4K via a pair of micro-HDMI ports, and hardware video decoding up to 4K p60, up to 4GB of RAM, two ports. 2.4/5.0 GHz band wireless networking, Bluetooth 5.0 capabilities, Gigabit Ethernet, USB 3.0, and PoE.

10.3.1.2 Camera Pi V2

To capture shots with Raspberry Pi, the new camera module v2 features a Sony IMX219 fixed focus sensor, for a high native resolution. Useful for time-lapse activities, motion detection, or security camera, the module is plugged into the CSI port of the Raspberry Pi. The parameters of the model of camera Pi V2: Fixed focus lens, native resolution of 8MP, 3280×2464 photo resolution, video resolution 1080p30, 720p60, and 640×480p90 Size: 25mm × 23mm × 9mm, Weight: 3g.

10.3.2 Software environment

10.3.2.1 Python

An interpretive, multi-paradigm, and multi-platform programming language. It favors programming imperatively functional and oriented structure. It is a language that can be used in many contexts and adapt to all types of users thanks to its specialized libraries.

The main libraries used in Python for facial recognition are:

- **OpenCV** (Open-Source Computer Vision Library): Library for image and video processing and analysis in C/C++, Python, and Linux. This library offers manipulation and acquisition of videos, various utilities, image processing, image analysis, vision, shape recognition, and graphical interface.

- Face recognition: Is a process to recognize faces according to their photos or videos following an algorithm.

- **Imutils:** Series of practical functions to make basic image processing functions such as translation, rotation, resizing, matplotlib image display, and edge detection. It is much easier with OpenCV and Python.

- **Pickle:** Used to serialize and deserialize a python object structure. Any object in python can be decapitated in order to save it to disk. So, pickle is used to serialize the object before writing it to the file. The idea is that this character stream contains all the information needed to reconstruct the object in a python.

- **Raspbian:** An open operating system based on debian distribution, optimized for the Raspberry Pi hardware. It comes with more than 35000 packages of precompleted software packages for easy development.

10.3.3 Application process/steps

The face recognition system seeks to identify the human face. This face may change in appearance depending on the lighting and facial expression. To accomplish this computer task, the face recognition system performs three steps:

10.3.3.1 Dataset creation
The dataset contains the images of the people that the system wants to recognize. This dataset is divided into subdirectories so that each subdirectory contains the images of a single person and takes the name of the person as the name of the subdirectory. For our application we have created 20 subdirectories (10 include images of women and 10 include images of men), each subdirectories contains 1500 images of the same person taken from different angles. These images are of type jpg, size 640×480 and RGB color space.

10.3.3.2 Training part
The algorithm detects first the faces that are in the dataset. The facial images are adjusted to take into account the different positions of the face, the size, the lighting, and the grey levels of the images. Then comes the alignment process, which allows a correct localization of the facial features. Then comes the extraction of facial features, such as eyes, nose, mouth, etc... Finally, we store these features in a pickle file.

10.3.3.3 Recognition part
The goal of this part is to match the characteristics of the detected face by camera pi with the characteristics of the different faces created by the training file. Let's see now how it works for a single face detected since it is the same principle for several detected faces. If the characteristics of the detected face are identical to those of the file created by the training code, the algorithm draws a square around the face detected and displays the name corresponding

to that person (Figure 10.6) Otherwise, the face is unknown. This process is repeated until we stop the program. The figure (Figure 10.5) illustrates the stages of learning and testing algorithms.

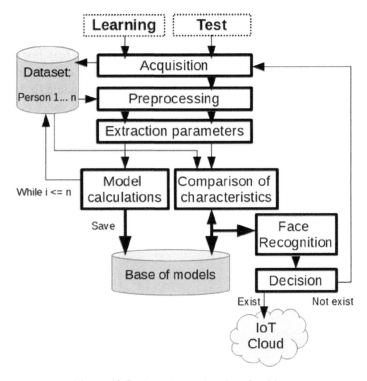

Figure 10.5 Learning and testing algorithm.

Figure 10.6 Detection and recognition test.

10.3.4 Implementation results and comparison

This program is applied for the three algorithms CNN, Hog, and Haar Cascade. To compare them, we have calculated each the rate and the speed of recognition. The (Table 10.1) below shows the results obtained.

Based on the results of the (Table 10.1) and the (Figure 10.7) we can conclude that Haar's cascade classifier is ahead in terms of implementation and speed, but shows more detection tiles (Figure 10.7(b), (c)). For face detection, the HOGs classifier is more accurate than the Haar cascade classifier, it can represent the local appearance very well, but it is unable to detect more

Table.10.1 Comparison of methods.

Rate of Recognition	State	CNN	HOG	Haar-cascade
	True	92%	85%	84%
	False	7.03%	2%	6.5%
	None	0.97%	13%	9.5%
	FPS	1.49	2.25	5.77
	Time	6h39min	3h41min	3h26min

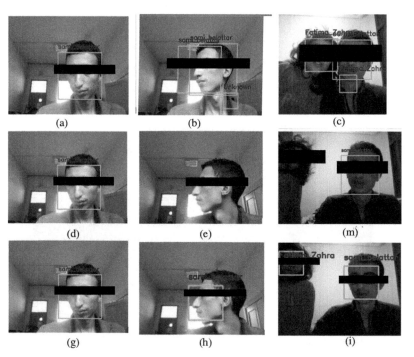

Figure 10.7 Result of face recognition using (a), (b), (c) haar cascade classifier, (d), (e), (f) hog classifier, (g), (h), (i) CNN classifier.

than one person (Figure 10.7(f)). Haar cascades and hog are faster but do not detect in different angles of view, and are less accurate (Figure 10.7(b), (e)). While CNNs are more accurate, and able to detect several people at once and in various positions (see Figure 10.7(i), (h)), but is slower. Haar cascade classifier does more false positive prediction on faces than the HOG classifier.

Therefore, CNN requires a graphics processing unit (GPU) with massive memory because they have heavy computations, but for devices like Raspberry Pi (4GB RAM), you should limit yourself to a shallower CNN. Unfortunately, the shallower CNNs do not reach a very high level of accuracy. For a very large training data set, it is better to use CNNs. But for the Raspberry Pi, the HOG classifier is the most suitable. The table below (Table 10.2) summarizes the advantages and disadvantages of each algorithm:

Algorithm	Strength Points	Weakness Points
CNN	• Highly accurate • Robust. • Detect faces form different angles of view. • Not require lighting conditions or bright-ness. • Detect multiple people at once.	• Need a heigh performance computer. • Not popular.
HOG	• Very fast. • More accurate than Haar-cascade.	• Face detection problem when changing angle. • Do not recognise mora than one person.
HAAR CASCADE	• Very fast.	• Give a lot of false result. • Less accurate.

10.4 Conclusion

In this chapter, we have developed a simple and effective face authentication system aimed at addressing surveillance and security issues. To realize this system, we used at the hardware level a Raspberry Pi4 electronic board, and a Camera Pi V2. At the software level, we used the programming language Python and the operating system Rasbian. The application starts with the creation of the database. Then comes the step of extracting the features of each image in the dataset. In the end, follows the step of comparing these features with the features of the face detected by the pi camera. For this code, we used three facial recognition algorithms namely hog, haar cascade, and CNN. After most, we compared the results of the facial recognition of these three algorithms.

References

[1] Adiono, T., Prakoso, K. S., Putratama, C. D., Yuwono, B., and Fuada, S. (2018). Hog-adaboost implementationfor human detection employing fpga altera de2-115.International Journal of Advanced Computer Scienceand Applications, 9(10):353–8.

[2] Albelwi, S. and Mahmood, A. (2017). A framework for de-signing the architectures of deep convolutional neuralnetworks. Entropy, 19(6):242.

[3] Balogh, Z., Magdin, M., and Molnár, G. (2019). Motiondetection and face recognition using raspberry pi, asa part of, the internet of things. Acta PolytechnicaHungarica, 16(3):167–185.

[4] Bruce, V. and Young, A. (1986). Understanding face recognition. British journal of psychology, 77(3):305–327.

[5] Dhiman, G. et al. (2020). An innovative approach for face recognition using raspberry pi. Artificial Intelligence Evolution, pages 102–107.

[6] Gauswami, M. H. and Trivedi, K. R. (2018). Implementation of machine learning for gender detection using cnn on raspberry pi platform. In2018 2nd International Conference on Inventive Systems and Control (ICISC), pages 608–613. IEEE.

[7] Gupta, I., Patil, V., Kadam, C., and Dumbre, S. (2016). Face detection and recognition using raspberry pi. In2016IEEE international WIE conference on electrical and computer engineering (WIECON-ECE), pages 83–86. IEEE.

[8] Mane, S. and Shah, G. (2019). Facial recognition, expression recognition, and gender identification. In Data Management, Analytics and Innovation, pages 275–290. Springer.

[9] Nikisins, O., Fuksis, R., Kadikis, A., and Greitans, M. (2015). Face recognition system on raspberry pi. In-stitute of Electronics and Computer Science, 14.

[10] Nord, J. H., Koohang, A., and Paliszkiewicz, J. (2019). The internet of things: Review and theoretical framework. Expert Systems with Applications, 133:97–108.

[11] Patnaik Patnaikuni, D. R. (2017). A comparative study of arduino, raspberry pi and esp8266 as iot development board. International Journal of Advanced Research in Computer Science, 8(5).

[12] Radzi, S. A., Alif, M. M. F., Athirah, Y. N., Jaafar, A., Nori-han, A., and Saleha, M. (2020). Iot based facial recognition door access control home security system using raspberry pi. International Journal of Power Electron-ics and Drive Systems, 11(1):417.

[13] Sajjad, M., Nasir, M., Muhammad, K., Khan, S., Jan,Z., Sangaiah, A. K., Elhoseny, M., and Baik, S. W.(2020).Raspberry pi assisted face recognition framework for enhanced law-enforcement services in smart cities. Future Generation Computer Systems,108:995–1007.

[14] Soldera, J., Schu, G., Schardosim, L. R., and Beltrao, E. T. (2017). Facial biometrics and applications. IEEE Instrumentation & Measurement Magazine, 20(2):4–30.

[15] Viola, P. and Jones, M. (2001). Rapid object detection us-ing a boosted cascade of simple features. In Proceedings of the 2001 IEEE computer society conference on computer vision and pattern recognition. CVPR 2001, volume 1, pages I–I. Ieee.

[16] Wazwaz, A. A., Herbawi, A. O., Teeti, M. J., and Hmeed,S. Y. (2018). Raspberry pi and computers-based face detection and recognition system. In2018 4th International Conference on Computer and Technology Ap-plications (ICCTA), pages 171–174. IEEE

SECTION 5

Internet of Things Based Embedded System

11

Survey Review on Artificial Intelligence and Embedded Systems for Agriculture Safety: A proposed IoT Agro-meteorology System for Local Farmers in Morocco

A.A. Mana[1], A. Allouhi, A. Hamrani[2], A. Jamil[1],
K. Ouazzani[1] A. Barrahmoune[3], D.Daffa[4]

[1]Laboratory of innovative technologies, USMBA, FES, Morocco.
[2]Department of Mechanical and Materials Engineering, Florida International University, USA.
[3]School of East Asian Studies, University of Sheffield, United Kingdom.
[4]School of networks and telecommunications, Fes, Morocco.
Email : abdelali.mana@usmba.ac.ma

Abstract

Agriculture is crucial to human life. It is the main food provider, yet it remains prone to climate change and other challenges, notably in developing countries. Some of the most prominent challenges are related to the surveillance and monitoring of climate, water resources, and soil quality. The evolution of Artificial Intelligence (AI), embedded systems, and the Internet of Things (IoT) is undeniable. Their adoption at a large scale in agricultural activities offers a great opportunity to overcome many of these challenges. Unfortunately, their adoption in many countries is still limited, and farmers are still the most forgotten of this evolution towards agriculture 4.0.

In this context, IoT-embedded systems can play a major role as a technology mediator that mutually boosts the productivity of farmlands and promote A.I. in agricultural fields. A distinctive feature of this study is to review and investigate the most suitable solutions for rural eras and familial agriculture. To this end, the present chapter proposes a study and design of a low-cost local weather station for local farmers. The proposed solution is based on the IoT networking systems to establish a Human-Technology-Environment relationship. The model is flexible to cluster more sensors and intelligent algorithms, to overcome issues related to climate uncertainty, which presents a real obstacle in the way to integrating precision in the agricultural sector.

11.1 Introduction

All around the world, agriculture is dramatically confronted with serious challenges, such as climate change, water scarcity, and heavy dependence on fossil fuels [1]. furthermore, the COVID19 pandemic threatens the entire agricultural value chain[2]. Indeed, the impacts are intensely intertwined and the agro-food chain is the most considered by these issues[3].

However, many developing countries have been excluded from most of the benefits of sustainable development [4]. After the Kyoto Protocol and COP22, little seems to have changed; but the truth is that traditional energy sources (firewood, biomass waste, human, and animal traction) remain the main and often the only energy resources available to millions of rural families[5]. Therefore, new approaches are immediately required to overwhelm classic practices that consider farms and ecosystems as industrial entities [6].

In fact, we are experiencing a world with a growing population and millions of people going hungry. It will take all our knowledge and imagination to deal in an integrated way with the challenges of maintaining soil fertility, and water resources, eliminating pests and diseases that affect both crops and animals. Thus, sustainability and precision in agriculture must focus on integrate environmental health, energetic profitability, social and economic fairness[7].

Actually, the global temperatures were estimated at 1.7 degrees surpassing the global average by approximately 0.15 degrees [8]. In a Mediterranean context, the dry and warm summer climate can influence agriculture. Those countries are highly exposed to climate change too, an increase in the degradation of lands is predicted as a consequence of reduced precipitation [9]. Thus, climate data remains one of the sources of information necessary for the integration of any possible technology related to the biosphere. In agriculture as for renewable energies, parameters such as wind, temperature, relative humidity, and availability of resources must be quantified and calculated rigorously in order to ensure a good prediction or use.

From another point of view, intelligent systems are able to process information, deliver complex reports, serve farmers in decision making and comply with quality requirements. Accordingly, AI has the potential to provide an essential solution to face different challenges and will make it possible to produce better and more, with fewer inputs. The diverse forms in which new technologies can intervene in agriculture operates through the use of embedded systems, the internet of things, and intelligent algorithms.

Thus, the chapter elaborates a survey review on the use of artificial intelligence and embedded systems in farming. As a consequence, the Internet of Things will be proposed as a reliable solution for sufficient agricultural precision and data collecting.

Considering the literature, several solutions have been suggested, and various models are being developed for IoT-based systems for smart farming. But, most of the stations are not committed to small farmers. Particularly, in developing countries, where internet access is even too limited. Recent systems aim to provide models to the researchers, instead of farmers.

In this context, an IoT Embedded System was proposed to connect marginalized and familial agriculture. It consists of offline access to data. The agro-weather station is designed to serve farmers, and engineers at once.

In the design of this platform, many sensors are used to retrieve information on key parameters such as air temperature and humidity, soil moisture, the presence of rain, and day/night-time detection. The recovered data will then be sent via a wi-fi Shield to the firebase. A user-friendly mobile application was developed using ionic framework to enable farmers to access this data and monitor their fields in real-time using a smartphone. In a regional prospect, this platform will offer the opportunity to facilitate the collection and access to information, and ease the management and monitoring of resources.

From the point-view of this chapter, to unlock the potential of embedded systems to digitalize agriculture, and promote the spreading of intelligence models and machines, it is primordial to integrate farmers into the heart of this transition. Managing information is the key element to achieving more sustainability and precision in agriculture.

Vast sequential data collection will drive the use of machine learning and AI and new models will need to be developed to make the data useful. Thus, the increased concern for big data and connected embedded systems, will drive the use of new and deep models of AI and build precise agricultural decisions.

However, countries that rely on agriculture for their economies must have the advantage of moving beyond old techniques that depend on climatic hazards and achieve sustainable, smart, and secure models in all stages of production.

11.2 AI-enabled Embedded Systems for Agriculture

11.2.1 Precision in water management

Agriculture is still an intensive consumer of water resources, with approximately 85% of freshwater available in the world [10]. A percentage that continues to increase in the face of demographic growth and food needs, which require more efficient solutions, in order to rationalize and manage water resources. Thus, the emergency to adopt new technologies in agriculture arose from serious constraints like water scarcity, increased food demand, and degradation of natural ecosystems.

To this end, technologies have been developed to facilitate the linking and data-sharing between the machines and different nodes installed in the agricultural field. We are talking about Machine to Machine (M2M) technology, capable of detecting moisture content and temperature at periodic intervals, in order to automate the irrigation of agricultural plots with exact needs [11]. Remote sensors using Arduino technology can be utilized in automation and irrigation monitoring [12] [13].

Otherwise, machine learning becomes an important tool in making a decision and mitigating water management problems up to a certain level, and sustaining the evapotranspiration process (Figure 11.1). At the same time, as reviewed in Table 11.1, ML allows for high accuracy and effective use of resources. One of the essential aspects of employment ML is the large-scale

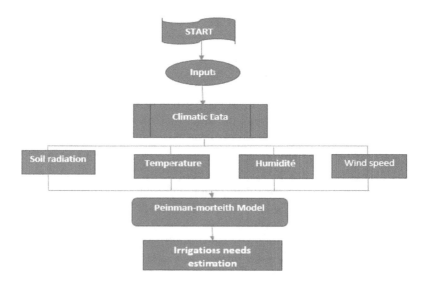

Figure 11.1   Process methodology to estimation plant water requirements [17].

Table 11.1 Literature survey on water management models.

Application	Inputs	Algorithms	Method Model	Technology	Performance
[18]	climatic conditions, soil moisture content	Partial Least Square Regression (PLSR)	Evapotranspiration model	Economic Hardware, sensors, (IOT).	increased efficiency and economic feasibility
[19]	temperature, soil humidity	Fuzzy Logic Controller	Penman–Monteith model	Wireless sensors	automated drip irrigation
[20]	climatic conditions, soil moisture content	ANN method	soil moisture model		precision and robustness of soil moisture prediction, water saving
[21]	soil water content and meteorological data	convolutional neural network	Pearson correlation, soil water content autocorrelation	deep learning	prediction accuracy 93%
[22]	moisture, weather forecast and water level	Machine Learning algorithm –	—	Near-infrared (NIR) spectroscopy IOT, ZigBee technology, Arduino microcontroller	Drought prediction
[23]	meteorological data, soil humidity.	artificial neural network (ANN) Machine learning techniques	Evapotranspiration model	unmanned aerial vehicle (UAV) IoT.	Performance to predict water stress
[24][25]	Different conditions	ANN	Optimal model sizing	hybrid intelligent systems	Sizing of optimal stand-alone PV-systems
[26]	Climatic Data	Regression comparison	Optimal model sizing	Standalone power supply (SAPS)	seasonal variability of solar insolation
[27]	Different conditions and imputes	genetic algorithms	Optimal model sizing	sizing grid-connected PV-system	stability voltage distribution
[28]	Climatic data	Feed Forward Neural Network Adaptive Neuro Fuzzy Inference	Optimal model sizing	Pumping systems	Photovoltaic power forecast

use of wireless sensor networks and IoT. In other hands, the development of thermal imaging in crops has industrialized thermal cameras that can offer new opportunities for estimating the hydraulic conditions of plants by acquiring thermal indices of plants, which help to precisely determine water needs [14]. In more advanced cases, AI reasoning of soil water balance and forecasting are able to optimize irrigation and secure farms against probable flooding, droughts, and disasters [15][16].

11.2.2 Integrated food safety

It is reported that the development of crops is highly susceptible to weeds, disease, or infestation by pests and insects. Furthermore, these conditions negatively influence crops and agricultural production. Thus increasing food safety is decisive [29]. As such, the precise and rapid detection of diseases or weeds will make it possible to take control measures and secure management of the fields [30]. In recent years, intelligent models are applied in safety crops applications, through insect pest monitoring, weed detection, and plant disease identification [31][32] [33].

Deep learning has lastly emerged with big data and visual technologies[34]. CNNs are the easiest type of DL to process images with less errors but depend on more accumulated experts datasets (Figure 11.2). Deep CNNs have gained the interest of researchers in Smart integrated management. Deep learning-based CNN, coupled with remote sensing, big data are more rapid and reliable. in recent advances, Feature Pyramid Network (FPN) are able to detect tiny objects through building and training a large amount of data[35][36]. In general, as perceived in Table 11.2, new technologies and AI methods have widely performed multi classification in an agricultural field.

11.2.3 Crop productivity and fertility

In many countries, precision agriculture is still referred to as satellite agriculture or site-specific crop management, because it uses satellite and aerial imagery, climatic forecasting, prediction application, and productivity indicators. By collecting those parameters, AI could emerge agro-technologies and increase crop profitability. ML makes it possible by learning from experience, analyze data from both inputs and outputs, and performing crop production with an enhanced high degree of precision [49] (Figure 11.2).

Therefore, using new models can solve crop health problems or nutrient deficiencies in the soil [51]. AI has the potential to examine phytosanitary models in agriculture and to better manage soil health, and fertilizers

Table 11.2 A.I performance and limitations for weed and disease detection.

Application	Inputs	Method/algorithms	Performance	limitations	Ref
Weed detection	UAV images	Fully Convolutional Network (FCN) method	weed mapping: 94% weed recognition: 88%	requires vast human expertise	[37]
Weed detection	hyper spectral images	ANNs, GA	Performance. Reduces trial and error.	Requires big data, expensive	[38]
Weed detection	hyper spectral images	SVM ANN	Quick detection.	Limited crops Only detects low levels of nitrogen weeds	[39]
Weed detection	Digital Image	DIA, GPS	accepted accuracy and success rate.	time response.	[40]
Weed prevention	Sensors and GPS data	ROBOTs. Sensor machine learning	Saves time and removes resistant weeds.	Expensive and affect soils	[41]
Weed prevention	Yield sensing and imagery data	Colour Based and Texture Based algorithms;	high accuracy 92.9%	expensive	[42]
Weed detection	(RGB)/ hyper spectral images	Deep Convolutional neural networks	High accuracy 98.23%	Requires big data.	[43]
Disease detection	UAV images (RGB)	CNN	Overall accuracy 89%, 94%	Requires big data and human expertise	[44]
Disease detection	UAS images	GA, ANN	Accurate results in the tested environment.	Inefficacy of DB in large scales.	[45]
	Data Base (DB)	Rule-Based Expert		Can affects good species	
Disease detection	UAV image and sensing data	Phenotyping technology, remote sensing methods	Early season detection and performance	Require big data	[46]
Disease detection	meteorological data	Fuzzy Logic (FL), Web GIS	High performance of forecasting	Internet dependence	[47]
Disease detection	DB	Web-Based Expert System	High performance	Internet dependence	[48]

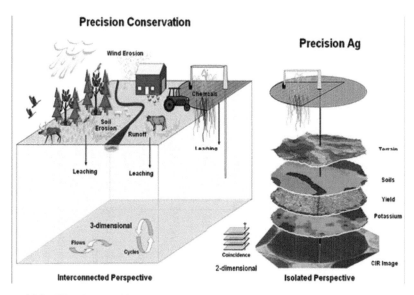

Figure 11.2   The site-specific crop management based on three-dimensional approach that assesses inputs and outputs from fields to watershed and regional scales [50].

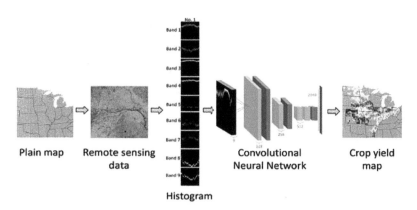

Figure 11.3   Crop yielding map using machine intelligence algorithms [54].

quantities [52]. By using AI, the chances of plant or soil degradation are reduced and crops are able to meet the market trends, maximize the return of different soils [53], and ensure a better crop mapping for decision-making (Figure 11.3).

11.2.4 Automation: Unmanned aerial vehicles (UAVs) and robots

In the current era, developing nations are implementing agricultural precision to increase crop profitability [55]. UAVs equipped with technologies

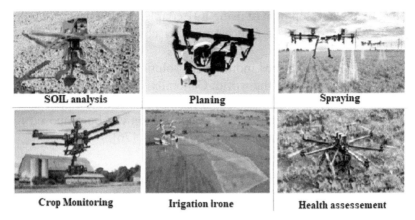

Figure 11.4 Drone applications in agriculture [71].

like cameras, servers, programmable interfaces, multi rotors, and automatic flights seem to be cost-effective in agricultural fields. Thus, using UAVs provides better image resolution compared to satellite confines [56].

There are many uses for drones in poultry farming such as crop monitoring [57], fertilizer spraying [58], crop height estimation [59], soil salinity management [60], seed planting [61], forest plantations, and biomass estimation [62]. Generally, The UAV remote sensing system is able for monitoring the temporal changes, backing for decision-making, improve land harvests, and expand economic profitability and systems cost-effectiveness [63]. Recently, the price of agricultural drones has progressively decreased, compared to satellites, and robots. [64], which can lead drones to promote farming management and improve more flexibility in agricultural practices. As shown in Figure 11.4. Drones can be used for several applications in Agriculture.

Now, further technology has obtained attention concerning automation through robotic systems.

Therefore, various recent works have been done to address agricultural problems using robotics for spraying fertilizers and pesticides [65], precision [66], identification [67], harvesting [68], and greenhouse farming [69][70].

11.2.5 Weather predictive analysis

Weather forecasts are essential to ensure the good progress of several agricultural activities. Also, when integrating renewable energies within the agriculture sphere, there is a vital requirement to gather the instantaneous values of these conditions. However, conventional methods provide hourly forecasts for large areas, which is often imprecise. In this context, forecasting can be expelled by the development of using IOT coupled with sensors [72]. Several solutions were proposed to forecast weather conditions in particular areas [73][74][75].

Table 11.3 summarizes more recent solutions for forecasting weather variables in many parts of the world. Generally, the design involves sensors coupled to a master controller. The gathered outputs are processed by a CPU or an online cloud platform. A mobile application can be proposed for the display of results with precision, in the short term [76].

11.3 Proposed Solution for Familial Agriculture and Small Farmers

11.3.1 Description of the study area

In this research, the location of the tests is delimited by the cities Meknes and Fez. Located in the central north of Morocco (Figure 11.5). Due to its history and its geographical location in the heart of the kingdom, the region of Fez-Meknes constitutes a strategic crossroads for various economic activities and for the internal and external animation of trade, divided between the provinces and prefectures [86]. The Fes-Meknes region is branded by rich soils and significant agricultural abilities. The climate in this region is considered semi-arid [87]. It should be noted that familial agriculture is the most expended in the region.

Investigations and calculation show that annual precipitation is about 475 mm, with important inter-monthly variations as perceived in Figure 11.6. While the average daily evapotranspiration is 7 mm. It is maximal during July and August and minimal during the months of December and January [17].

Actually, the agricultural area in the Fes-Meknes region represents 15% of the national useful area with approximately 1,335,639 hectares. While only 9% of these lands is irrigated [88]. The region thus benefits from a large arable land (about 1.4 million ha), a favourable climate, significant quantitative human resources: more than 1.7 million Rural areas, and qualitatively: this human potential is renowned for know-how various branches of the sector (traditional agro-sylvo-pastoral systems, mobilization of water resources ...), pedoclimatic conditions generally favorable for the mobilization large areas for cereal growing and vegetable crops, rich soils, with high productive potential, notably the Sais plateau, which explains the irrigation by forage is the most answered in the region. Another pragmatic level is the definition of the drilling point.

Table 11.3 Performances and limitations of A.I on weather predictability.

Applications	Technologies	Imputes	Performance	Limitations	Reference
Weather forecasting	Wireless sensors network, fuzzy control	Air (T, atm and RH), rainfall, radiation, and wind speed and direction climatic data	Meteorological disaster early warning, Forecasting	Need of connection in farmland, debugging process	[77]
Greenhouse monitoring	n ZigBee and GPRS wireless network	temperature, humidity, light intensity, and rainfall	real-time detection of greenhouse environmental factor	——	[78]
Farm facilities environment	wireless sensors network, GPRS	Crop storage temperature and moisture levels	Cost feasibility, helps farmers to achieve a more quality of crops	Short battery lifetime	[79]
Forecasting	Machine learning models, ANN	metrological parameters	Forecasting min and max temperature, long distance communication	Short period prediction (10 days)	[80]
Crops Water management	Machine learning, IoT	Crop proprieties, and precipitation	autonomous irrigation, low error rate	Cost	[81]
Smart farming	IoT	soil quality, environmental conditions fertilisation, and irrigation data	Elasticity and scalability of the platform		[82]
Weather forecasting	ANNs	Metrological parameters	Acceptable errors percentage, fast prediction. prediction results till 2050	Increasing of percentage predicted errors with time.	[83]
Load forecasting	ML,IoT	Different parameters	Acceptable errors, Fast prediction.	Short load prediction	[84]
Load forecasting	Multi Linear Regression (MLR)	Different parameters	higher accuracy	Short term	[85]

Figure 11.5   Position of the Fes-Meknes region in Morocco territorial division.

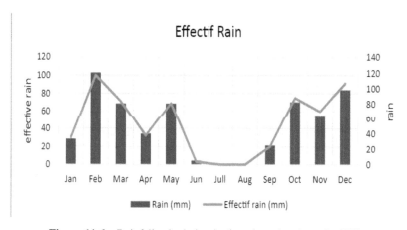

Figure 11.6   Rainfall calculation in the selected main region[17].

11.3.2 Model architecture

This section involves the design and implementation of a low-cost weather station platform for local farmers. In the design of our platform, we have sensors that can retrieve information on temperature and humidity, soil moisture, rain, and detection of light. Bits of information are transferred to the opensource firebase via a wi-fi shield. A developed graphical application with Ionic guarantees real-time access to data (see Figure 11.7).

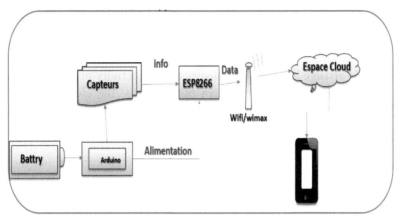

Figure 11.7 Functional diagram of the proposed solution.

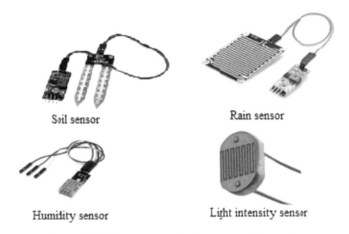

Figure 11.8 Sensors used in the agro-weather station.

11.3.3 Wireless sensors networks for agricultural Forecasting

In this part, we deal with the undeniable hardware and software elements in the agro-weather station. The system is flexible and can join more sensors able to collect more parameters and conditions in real-time with high accuracy. Figure 11.8 shows multi-sensors that can be clustered to the Arduino master card.

• **Master board control**: The role of the Arduino card is to store programs and ensure the employment of multiple sensors. For networking the system, the board is used as an expansion card with various functions such as: relays, motor controls, SD card readers, ethernet, wi-fi, GSM, GPS, bluetooth, clock, and LCD displays with the sensors [89].

- **Soil Moisture Sensor:** The electrical conductivity of the earth depends on the humidity of the soil. In other words, the electrical resistance of soil increases with the drought of this one. To measure this electrical resistance, two electrodes are used which are fixed on fork-shaped support that is planted vertically in the ground. It has two outputs: one digital with an adjustable threshold by a potentiometer and the other devices. The YL 69 sensor is a simple humidity sensor that can be used to detect when soil is lacking in water (high level) or not (low level). The sensor can be widely used for Automatic watering of indoor plants, Watering the garden, Irrigation of crops, Analog humidity measurements, Flood alarm, and Rain detectors [90].

- **Temperature and humidity sensor:** DHT11 Sensor collect mutually temperature with $\pm 2°$, and humidity with 5% precision. First, the Arduino activates the sensor by placing the data line at LOW for at least 800µs). During this time, the sensor prepares the temperature and humidity measurement. Once the data line is released to the sensor responds to the microcontroller by keeping the data line at LOW for 80µs, then to HIGH for 80µs. The sensor will then transmit a series of 40 bits (5 bytes). Both first bytes contain the humidity measurement, and the next two bytes contain the temperature measurement; while the fifth byte contains a sum control which verifies that the data is correct. The code for testing sensor and variations of air temperature and humidity in real-time are shown in (Appendix A).

- **Precipitation Sensor:** It includes a printed circuit board (control board), which "collects" the drops of rain. When the raindrops are collected on the circuit board, they create parallel resistance paths which are measured through the amplifier operational. The lower the resistance (or the more water), the higher the output voltage is low. Conversely, the less water there is, the higher the output voltage on the spindle. This sensor has 2 outputs: - a logic output D0 which allows an all or nothing detection (a screw is used to adjust the detection threshold). This output is at 1 when the detection plate is dry, and at 0 when it is wet. An analog output A0, which varies from 0V (completely wet plate) to 5V (dry plate).

After having carried out the assembly, to test the MH-RD the Arduino project can use the code source presented in (Appendix B).

- **Light sensor:** A photo-resistor or LDR is a dependent resistor sensor to light. In a decreased light, the sensor resistance increases respectively. This characteristic makes the LDR more useful for greenhouses and automated solar pumping.

11.3.4 Communication modules

- **Module RTC DS3231:** Communication of the RTC is shaped through the I2C interface to send and receive pieces of data. It is necessary to get the DATE and TIME information through this interface. The module can operate on a regulated supply of + 5V. Then the module's SDA links to the controller SDA and SCL are connected to the controller's SCL. Usually, the exchange is made byte by byte. When the current drops, the RTC module is automatically powered from the battery source. Time will be up to date, and when the system restarts, the controller is updated with no errors (Appendix C).

- **Networking system:** ESP8266 WI FI unit can be well-ordered from the internet. The ESP8266 communicates using a sequential interface. It uses Arduino's Rx and Tx pins connected to the ESP8266-12E to receive orders and communicate again. The module offers a full TCP/ UDP load and can be arranged as a web server. Thus, to send data to firebase, the ESP must be connected to the router.

- **Data-base and Mobile application:** As the access to the platform in real-time is a requirement, we have developed a mobile application that gives the user access to the information that the platform provides in real-time. In this work, we were interested in firebase technologies which provide the database where online information is stored, and ionic which allows us to develop a single application executable on all mobile platforms (Android, IOS, Windows phone, etc.). To send data to firebase, it is necessary to connect the ESP to the router. Then, install the package ArduinoFirebase et ArduinoJson, before using the code source to manage to send data from the embedded system to firebase. The ESP8266-12E Node MCU Kit is the module we used to establish the connection with the firebase database and also to transmit the data returned by the various sensors (Appendix D).

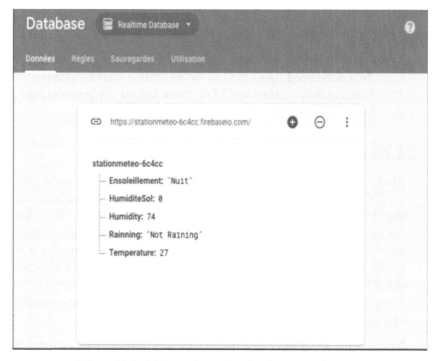

Figure 11.9   Main results retrieved and stored on firebase.

Indeed, pieces of information can be very structured (relational database for example), or hosted in the form of unstructured raw data (NoSQL database) which is the case of the database used for the storage of our data. information retrieved (Figure 11.8).

Mobile application is an application software developed for a mobile electronic device, such as a personal assistant, a cell phone, a [91]. Phones use aims to improve monitoring productivity and facilitate the retrieval of information such as email, electronic calendar, contacts, stock market, and weather information. In our case, Ionic technology is used to develop the display application. Once connected, this last will allow real-time access to the information stored in the firebase database. While Figure 11.9, shows the mockup of mounting the sensors with the Arduino board. The various Arduino boards will be powered by a rechargeable battery. But for future needs, it will be a question of designing an autonomous power supply system.

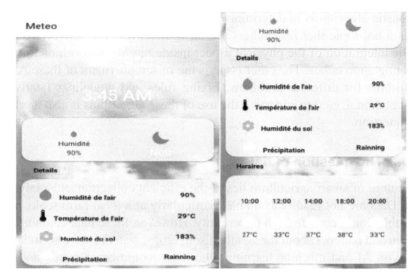

Figure 11.10 Mobile application interface.

11.4 Discussions: Questions and Challenges Raised by the use of AI and IoT in Agriculture

Sensors technique has completely revolutionized agriculture in the last years. However, the questions and challenges rising in using and exploiting AI in agriculture in the present and future time are not clear. The most relevant questions are cited in the following.

11.4.1 The question of trust

Trusting AI in the decision-making process for our food production has its advantages and drawbacks. The advantages are clearly visible in the wide embracing of this technology for automation processes and taking decisions based on data. The limitations of trusting these algorithms in agriculture can be seen as the emerging self-driving car technology. It will be difficult to define ethics for insurance to cover liability in case of fatal decisions and accidents. The need for normative rules for these innovations is the next step in the near future.

11.4.2 The question of applying AI stochastic algorithms

Data used by AI/ML to model complex living ecosystems such as environment, soils, and crops, lose all the physics behind it. Integrating these stochastic algorithms in deterministic approaches such as biophysical models is a hot topic that researchers are on it. Scientists are actively working on the integration of the physical aspect inside the ML algorithms for more targeting approaches. The other issue is the misemployment of these AI/ML algorithms for different problems. Fixing rules and guidelines (statistical, computational, etc.) to supervise the use of these algorithms is also a primordial question.

11.4.3 The question of data

The future of smart agriculture lies in the efficient collection and analysis of data. Data are not readily available, particularly at a local farm scale, and if available could contain high uncertainty. However, these data contain information and patterns about the weather, soil, crops, water resources, and more, that with AI and machine learning algorithms could be extracted and used as decision support for farmers, researchers, agricultural advisors, market services, and input suppliers. Development of an open-source database, at a global and local scale, will serve as a baseline for scientists, economists, and farmers. This database will contain satellite imagery, Internet of Things (IoT) sensors data, weather forecasting, and other data concerning soil, crops, water, tillage and surface temperature, etc. At the stage of data collection, the IoT network can help to collect data measured from sensors located in the field, in the soil, in a tower, or mobile on tractors, and make them available in real-time. The next stage concerns the integration of collected information with other data from cloud-based systems such as databases on crop and soil types, present and future weather conditions, and cost models to finally extract insights and patterns by machine learning models. These predictive models assist farmers and scientists to detect existing and future issues. The challenge now lies in promoting global efforts for the availability, accessibility, and usability of data in agriculture [92].

11.4.4 The question of interpretability

AI and ML models, howsoever powerful they may be, they stay considered mysterious and black boxes. At the present time, it is difficult to measure and justify their results. Interpretability is the inherent issue with the use of AI. Two levels of interpretability can be distinguished in machine learning models:

1. high interpretability: this level includes basically classical regression algorithms such as linear, multiple linear, decision trees, ridge, and least absolute shrinkage and selection operator (LASSO) regressions.

2. Low interpretability: this includes ML models such as support vector machine (SVM), neural networks, and deep learning. The lake of interpretability is justified by the use of multiple interconnected layers' structures containing different types of neurons; in the case of deep learning, or complex geometrical foundations; in the case of SVM.

11.5 Conclusion and Future Works

The processes and methods of production in agriculture have altered dramatically during and after the COVID-19 pandemic. Agriculture was hampered by labor shortages and resource disruptions. Besides, the impact of the COVID-19 pandemic on smart agricultural operations is a source of worry. The vulnerability of agriculture, for example, has risen, prompting each country to refocus on food safety and work to enhance food production. The evolution of agriculture towards the new agro food 4.0 will encourage businesses and farmers to invest in automation and artificial intelligence.

Thus, the current chapter proposes the use of new technologies to support familial agriculture. Importantly, it provides an in-depth understanding of promising different applications for agriculture. Therefore, the positive impacts of the use of new technologies in agriculture include:

*   AI methods are more robust for detection, analysis, and estimation by combining IoT with CNN, RNN, or any other computational networks.

*   Wireless technology and IOT may use the latest communication protocols and sensors are more favorable for forecasting in agricultural fields.

*   Distant meteorological monitoring and control of crops can be valuable for small farmers.

Furthermore, the proposed agro-weather station is one of the many aspects of using the computer tool coupled with electronics.

Indeed, we strongly hope that through the simplification of the solution, it can serve as a basic element for other more in-depth studies to integrate it into more complex and large-scale systems. But, above all, to make the rural farmer at the heart of this technological transition, by offering a low-cost solution, which can be consulted in real-time.

Obviously, it is strongly recommended to improve the ability and the robustness of the system. In future work, it will be interesting to cluster, the IoT station with intelligent algorithms for dedicated greenhouses. The period of collecting DATA will be extended from days to months. Thus greenhouse

Indoor-conditions predictions can serve to manage irrigation, ventilation, CO2 injection, and plant needs. Recurrent Neural Networks can be selected as the most suitable model to collect sequential DATA as input, to train and predict future information.

Open AI sources can be used, like Tensorflow and Keras, to test and train vast data collected. Different intelligent models can be clustered to the proposed platform to assist different end-users' roles like farmers, researchers, and engineers.

Otherwise, the proposed station can add more sensors and actuators to automate farming activities.

In general, new technologies must combine intelligent models with sustainability and social challenges to promote farming activities like planting, spraying, and even harvesting. However, digital qualification and IT literacy are still low in several countries, notably in rural areas. But we think that low-cost solutions, and the expanding use of smartphones can serve as momentum to transit from familial to sustainable agriculture.

To conclude, this chapter would certainly asset the research community concerned with smart farming, as well as engineers developing applications for automation.

Appendix A

```
1    #include <dht.h>
2
3    dht DHT;
4
5    #define DHT11_PIN 7
6    void setup() {
7        Serial.begin(9600);
8    }
9
10   void loop() {
11     int chk = DHT.read11(DHT11_PIN);
12     Serial.print("Temperature = ");
13     Serial.println(DHT.temperature);
14     Serial.print("Humidity = ");
15     Serial.println(DHT.humidity);
16     delay(1000);
17   }
```

Program to test the DHT11 sensor.

Test result with DHT11.

Appendix B

```
1    //const int capteur_D = 4; // pin connected to the digital output of the sensor
2    const int capteur_A = A0; // pin connected to the analog output of the sensor
3    int analog_val;
4    void setup()
5    {
6      pinMode(capteur_D, INPUT);
7      //pinMode(capteur_A, INPUT);
8      Serial.begin(9600);
9    }
10   void loop()
11   {
12   //Digital part
13
14   if(digitalRead(capteur_D) == LOW) // The sensor is active in the low state
15     {
16       Serial.println("Digital value : wet");
17       delay(10); // Tempo
18     }
19   else
20     {
21       Serial.println("Digital value : not wet");
22       delay(10); // Tempo
23     }
24   // Analog part
25   analog_val=analogRead(capteur_A); // we read the "analog" sensor pin
26   Serial.print("Analog value : ");
27   Serial.println(analog_val); // display the detection value on the serial link
28   Serial.println("");
29     delay(1000);
30   }
```

Program to test the MH-RD sensor.

Appendix C

First assembly to test precision of sensors used and RTC connexion.

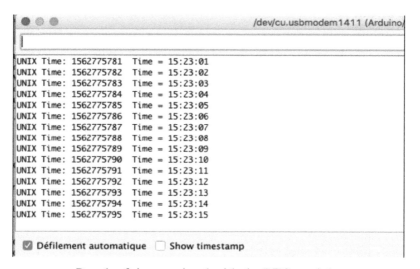

Result of time retrieval with the RTC module.

Appendix D

```
1    #include <ESP8266WiFi.h>
2    #include <FirebaseArduino.h>
3    #include "DHT.h"
4    #define DHTPIN 13      // what digital pin we're connected to
5    #define DHTTYPE DHT11   // DHT 11
6    // Set these to run example.
7    #define FIREBASE_HOST "stationmeteo-b6ces.firebaseio.com"
8    #define FIREBASE_AUTH "HtxO3apkYimdNxtSSi6fBrSXVB5oBQi7cq7RcdQ"
9    #define WIFI_SSID "Last Batman"
10   #define WIFI_PASSWORD "#wyTeamNoLimits"
11
12   DHT dht(DHTPIN, DHTTYPE);
13   void setup() {
14      Serial.begin(9600);
15      Serial.println("DHTxx test!");
16      dht.begin();
17      // connect to wifi.
18      WiFi.begin(WIFI_SSID, WIFI_PASSWORD);
19      Serial.print("connecting");
20      while (WiFi.status() != WL_CONNECTED) {
21         Serial.print(".");
22         delay(500);
23      }
24      Serial.println();
25      Serial.print("connected: ");
26      Serial.println(WiFi.localIP());
27      Firebase.begin(FIREBASE_HOST, FIREBASE_AUTH);
28   }
29   void loop() {
30      delay(2000);
31
32      // Reading temperature or humidity takes about 250 milliseconds!
33      // Sensor readings may also be up to 2 seconds 'old' (its a very slow sensor)
34      float h = dht.readHumidity();
35      // Read temperature as Celsius (the default)
36      float t = dht.readTemperature();
37      Serial.print("Humidity: ");
38      Serial.print(h);
39      Serial.print(" %\t");
40      Serial.print("Temperature: ");
41      Serial.print(t);
42      Serial.print(" *C ");
43      // set temperature
44      Firebase.setFloat("Temperature", t);
45      // handle error
46      if (Firebase.failed()) {
47         Serial.print("setting /Temperature failed:");
48         Serial.println(Firebase.error());
49         return;
50      }
51      delay(1000);
52
53      // set humidity
54      Firebase.setFloat("Humidity", h);
55      // handle error
56      if (Firebase.failed()) {
57         Serial.print("setting /humidity failed:");
58         Serial.println(Firebase.error());
59         return;
60      }
61      delay(1000);*/
62
63      String temperature = Firebase.pushFloat("Temperature", t);
64      // handle error
65      if (Firebase.failed()) {
66         Serial.print("pushing /Temperature failed:");
67         Serial.println(Firebase.error());
68         return;
69      }
70      delay(1000);
71      Serial.println(temperature);
72
73      // set humidity
74      String humidity = Firebase.pushFloat("Humidity", h);
75      // handle error
76      if (Firebase.failed()) {
77         Serial.print("pushing /humidity failed:");
78         Serial.println(Firebase.error());
79         return;
80      }
81      delay(1000);
82      Serial.println(humidity);
83
84      // get temperature
85      Serial.print("Temperature: ");
86      Serial.println(Firebase.getFloat("Temperature"));
87      delay(1000);
88
89      // get humidity
90      Serial.print("Humidity: ");
91      Serial.println(Firebase.getFloat("Humidity"));
92      delay(1000);*/
```

Program to connect the system with firebase.

References

[1] H. Turral, J. Burke, and J.-M. Faurès, *Climate change, water and food security.*, no. 36. Food and Agriculture Organization of the United Nations (FAO), 2011.

[2] S. Aday and M. S. Aday, "Impact of COVID-19 on the food supply chain," *Food Qual. Saf.*, vol. 4, no. 4, pp. 167–180, 2020.

[3] F. FAO, "The future of food and agriculture–Trends and challenges," *Annu. Rep.*, 2017.

[4] L. T. Clausen and D. Rudolph, "Renewable energy for sustainable rural development: Synergies and mismatches," *Energy Policy*, vol. 138, p. 111289, 2020.

[5] C. Aall, K. Moberg, J.-P. Cerone, E. Reimerson, and F. Dorner, "Policies for reducing household green house gas emissions," 2017.

[6] A. A. Mana, A. Allouhi, K. Ouazzani, and A. Jamil, "Feasibility of agriculture biomass power generation in Morocco: Techno-economic analysis.," *J. Clean. Prod.*, p. 126293, 2021.

[7] J. L. Johnston, J. C. Fanzo, and B. Cogill, "Understanding sustainable diets: a descriptive analysis of the determinants and processes that influence diets and their impact on health, food security, and environmental sustainability," *Adv. Nutr.*, vol. 5, no. 4, pp. 418–429, 2014.

[8] H.-O. P. Mbow, A. Reisinger, J. Canadell, and P. O'Brien, "Special Report on climate change, desertification, land degradation, sustainable land management, food security, and greenhouse gas fluxes in terrestrial ecosystems (SR2)," *Ginevra, IPCC*, 2017.

[9] V. Simonneaux, A. Cheggour, C. Deschamps, F. Mouillot, O. Cerdan, and Y. Le Bissonnais, "Land use and climate change effects on soil erosion in a semi-arid mountainous watershed (High Atlas, Morocco)," *J. Arid Environ.*, vol. 122, pp. 64–75, 2015.

[10] N. Walmsley and G. Pearce, "Towards sustainable water resources management: bringing the Strategic Approach up-to-date," *Irrig. Drain. Syst.*, vol. 24, no. 3–4, pp. 191–203, 2010.

[11] Y. Shekhar, E. Dagur, S. Mishra, and S. Sankaranarayanan, "Intelligent IoT based automated irrigation system," *Int. J. Appl. Eng. Res.*, vol. 12, no. 18, pp. 7306–7320, 2017.

[12] J. Muangprathub, N. Boonnam, S. Kajornkasirat, N. Lekbangpong, A. Wanichsombat, and P. Nillaor, "IoT and agriculture data analysis for smart farm," *Comput. Electron. Agric.*, vol. 156, pp. 467–474, 2019.

[13] K. Jha, A. Doshi, P. Patel, and M. Shah, "A comprehensive review on automation in agriculture using artificial intelligence," *Artif. Intell. Agric.*, vol. 2, pp. 1–12, 2019.

[14] S. Fuentes, R. De Bei, J. Pech, and S. Tyerman, "Computational water stress indices obtained from thermal image analysis of grapevine canopies," *Irrig. Sci.*, vol. 30, no. 6, pp. 523–536, 2012.

[15] K. E. Adikari, S. Shrestha, D. T. Ratnayake, A. Budhathoki, S. Mohanasundaram, and M. N. Dailey, "Evaluation of artificial intelligence models for flood and drought forecasting in arid and tropical regions," *Environ. Model. Softw.*, vol. 144, p. 105136, 2021.

[16] F. A. Prodhan *et al.*, "Deep learning for monitoring agricultural drought in South Asia using remote sensing data," *Remote Sens.*, vol. 13, no. 9, p. 1715, 2021.

[17] A. A. Mana, A. Allouhi, K. Ouazzani, and A. Jamil, "Toward a Sustainable Agriculture in Morocco Based on Standalone PV Pumping Systems: A Comprehensive Approach," *Adv. Technol. Sol. Photovoltaics Energy Syst.*, p. 399.

[18] S. Choudhary, V. Gaurav, A. Singh, and S. Agarwal, "Autonomous Crop Irrigation System using Artificial Intelligence," *Int. J. Eng. Adv. Technol.*, vol. 8, no. 5S, pp. 46–51, 2019.

[19] K. Anand, C. Jayakumar, M. Muthu, and S. Amirneni, "Automatic drip irrigation system using fuzzy logic and mobile technology," in *2015 IEEE Technological Innovation in ICT for Agriculture and Rural Development (TIAR)*, 2015, pp. 54–58.

[20] M. S. P. Subathra, C. J. Blessing, S. T. George, A. Thomas, A. D. Raj, and V. Ewards, "Automated Intelligent Wireless Drip Irrigation Using ANN Techniques," in *Advances in Big Data and Cloud Computing*, Springer, 2019, pp. 555–568.

[21] H. Chen *et al.*, "A deep learning CNN architecture applied in smart near-infrared analysis of water pollution for agricultural irrigation resources," *Agric. Water Manag.*, vol. 240, p. 106303, 2020.

[22] G. Arvind, V. G. Athira, H. Haripriya, R. A. Rani, and S. Aravind, "Automated irrigation with advanced seed germination and pest control," in *2017 IEEE Technological Innovations in ICT for Agriculture and Rural Development (TIAR)*, 2017, pp. 64–67.

[23] T. Poblete, S. Ortega-Farías, M. A. Moreno, and M. Bardeen, "Artificial neural network to predict vine water status spatial variability using multispectral information obtained from an unmanned aerial vehicle (UAV)," *Sensors*, vol. 17, no. 11, p. 2488, 2017.

[24] A. Mellit, M. Menghanem, and M. Bendekhis, "Artificial neural network model for prediction solar radiation data: application for sizing stand-alone photovoltaic power system," in *IEEE Power Engineering Society General Meeting, 2005*, 2005, pp. 40–44.

[25] A. Mellit and M. Benghanem, "Sizing of stand-alone photovoltaic systems using neural network adaptive model," *Desalination*, vol. 209, no. 1–3, pp. 64–72, 2007.

[26] B. S. Richards and G. J. Conibeer, "A comparison of hydrogen storage technologies for solar-powered stand-alone power supplies: A photovoltaic system sizing approach," *Int. J. Hydrogen Energy*, vol. 32, no. 14, pp. 2712–2718, 2007.

[27] J. C. Hernández, A. Medina, and F. Jurado, "Optimal allocation and sizing for profitability and voltage enhancement of PV systems on feeders," *Renew. Energy*, vol. 32, no. 10, pp. 1768–1789, 2007.

[28] R. Ben Ammar, M. Ben Ammar, and A. Oualha, "Photovoltaic power forecast using empirical models and artificial intelligence approaches for water pumping systems," *Renew. Energy*, vol. 153, pp. 1016–1028, 2020.

[29] E. Skotadis *et al.*, "A sensing approach for automated and real-time pesticide detection in the scope of smart-farming," *Comput. Electron. Agric.*, vol. 178, p. 105759, 2020.

[30] C.-L. Chung, K.-J. Huang, S.-Y. Chen, M.-H. Lai, Y.-C. Chen, and Y.-F. Kuo, "Detecting Bakanae disease in rice seedlings by machine vision," *Comput. Electron. Agric.*, vol. 121, pp. 404–411, 2016.

[31] A. N. Alves, W. S. R. Souza, and D. L. Borges, "Cotton pests classification in field-based images using deep residual networks," *Comput. Electron. Agric.*, vol. 174, p. 105488, 2020.

[32] L. Nanni, G. Maguolo, and F. Pancino, "Insect pest image detection and recognition based on bio-inspired methods," *Ecol. Inform.*, vol. 57, p. 101089, 2020.

[33] S. Shrivastava, S. K. Singh, and D. S. Hooda, "Soybean plant foliar disease detection using image retrieval approaches," *Multimed. Tools Appl.*, vol. 76, no. 24, pp. 26647–26674, 2017.

[34] M. Zomeni, J. Tzanopoulos, and J. D. Pantis, "Historical analysis of landscape change using remote sensing techniques: An explanatory tool for agricultural transformation in Greek rural areas," *Landsc. Urban Plan.*, vol. 86, no. 1, pp. 38–46, 2008.

[35] S. Dong *et al.*, "CRA-Net: A channel recalibration feature pyramid network for detecting small pests," *Comput. Electron. Agric.*, vol. 191, p. 106518, 2021.

[36] R. Wang, L. Jiao, C. Xie, P. Chen, J. Du, and R. Li, "S-RPN: Sampling-balanced region proposal network for small crop pest detection," *Comput. Electron. Agric.*, vol. 187, p. 106290, 2021.

[37] H. Huang, J. Deng, Y. Lan, A. Yang, X. Deng, and L. Zhang, "A fully convolutional network for weed mapping of unmanned aerial vehicle (UAV) imagery," *PLoS One*, vol. 13, no. 4, p. e0196302, 2018.

[38] A. Tobal and S. A. Mokhtar, "Weeds identification using Evolutionary Artificial Intelligence Algorithm.," *J. Comput. Sci.*, vol. 10, no. 8, pp. 1355–1361, 2014.

[39] Y. Karimi, S. O. Prasher, R. M. Patel, and S. H. Kim, "Application of support vector machine technology for weed and nitrogen stress detection in corn," *Comput. Electron. Agric.*, vol. 51, no. 1–2, pp. 99–109, 2006.

[40] R. Gerhards and S. Christensen, "Real-time weed detection, decision making and patch spraying in maize, sugarbeet, winter wheat and winter barley," *Weed Res.*, vol. 43, no. 6, pp. 385–392, 2003.

[41] "Fighting weeds: Can we reduce, or even eliminate, herbicides by utilizing robotics and AI? | Genetic Literacy Project." https://geneticliteracyproject.org/2018/12/12/fighting-weeds-can-we-reduce-or-even-eliminate-herbicide-use-through-robotics-and-ai/ (accessed Nov. 29, 2020).

[42] M. Sujaritha, S. Annadurai, J. Satheeshkumar, S. K. Sharan, and L. Mahesh, "Weed detecting robot in sugarcane fields using fuzzy real time classifier," *Comput. Electron. Agric.*, vol. 134, pp. 160–171, 2017.

[43] M. H. Asad and A. Bais, "Weed detection in canola fields using maximum likelihood classification and deep convolutional neural network," *Inf. Process. Agric.*, 2019.

[44] M. D. Bah, A. Hafiane, and R. Canals, "Deep learning with unsupervised data labeling for weed detection in line crops in UAV images," *Remote Sens.*, vol. 10, no. 11, p. 1690, 2018.

[45] K. Balleda, D. Satyanvesh, N. Sampath, K. T. N. Varma, and P. K. Baruah, "Agpest: An efficient rule-based expert system to prevent pest diseases of rice & wheat crops," in *2014 IEEE 8th International Conference on Intelligent Systems and Control (ISCO)*, 2014, pp. 262–268.

[46] Y. Ampatzidis and V. Partel, "UAV-based high throughput phenotyping in citrus utilizing multispectral imaging and artificial intelligence," *Remote Sens.*, vol. 11, no. 4, p. 410, 2019.

[47] V. Tilva, J. Patel, and C. Bhatt, "Weather based plant diseases forecasting using fuzzy logic," in *2013 Nirma University International Conference on Engineering (NUiCONE)*, 2013, pp. 1–5.

[48] Z. Beiranvand, "Integration of expert system and fuzzy theory for diagnosis wheat plant diseases," *QUID Investig. Cienc. y Tecnol.*, no. 1, pp. 1924–1930, 2017.

[49] S. Y. Liu, "Artificial Intelligence (AI) in agriculture," *IEEE Comput. Archit. Lett.*, vol. 22, no. 03, pp. 14–15, 2020.

[50] J. Delgado, N. M. Short, D. P. Roberts, and B. Vandenberg, "Big Data Analysis for Sustainable Agriculture," *Front. Sustain. Food Syst.*, vol. 3, p. 54, 2019.

[51] A. Hamrani, A. Akbarzadeh, and C. A. Madramootoo, "Machine learning for predicting greenhouse gas emissions from agricultural soils," *Sci. Total Environ.*, vol. 741, p. 140338, 2020.

[52] A.-K. Mahlein, "Plant disease detection by imaging sensors–parallels and specific demands for precision agriculture and plant phenotyping," *Plant Dis.*, vol. 100, no. 2, pp. 241–251, 2016.

[53] D. I. Patrício and R. Rieder, "Computer vision and artificial intelligence in precision agriculture for grain crops: A systematic review," *Comput. Electron. Agric.*, vol. 153, pp. 69–81, 2018.

[54] "Crop yield analysis — Sustainability and artificial intelligence lab." http://sustain.stanford.edu/crop-yield-analysis (accessed Jan. 02, 2021).

[55] U. M. R. Mogili and B. Deepak, "Review on application of drone systems in precision agriculture," *Procedia Comput. Sci.*, vol. 133, pp. 502–509, 2018.

[56] H. Bu, L. K. Sharma, A. Denton, and D. W. Franzen, "Comparison of satellite imagery and ground-based active optical sensors as yield predictors in sugar beet, spring wheat, corn, and sunflower," *Agron. J.*, vol. 109, no. 1, pp. 299–308, 2017.

[57] J. Bendig, A. Bolten, and G. Bareth, "Introducing a low-cost mini-UAV for thermal-and multispectral-imaging," *Int. Arch. Photogramm. Remote Sens. Spat. Inf. Sci*, vol. 39, pp. 345–349, 2012.

[58] Y. Huang, W. C. Hoffmann, Y. Lan, W. Wu, and B. K. Fritz, "Development of a spray system for an unmanned aerial vehicle platform," *Appl. Eng. Agric.*, vol. 25, no. 6, pp. 803–809, 2009.

[59] D. Yu *et al.*, "Improvement of sugarcane yield estimation by assimilating UAV-derived plant height observations," *Eur. J. Agron.*, vol. 121, p. 126159, 2020.

[60] K. Ivushkin *et al.*, "UAV based soil salinity assessment of cropland," *Geoderma*, vol. 338, pp. 502–512, 2019.

[61] E. E. da Silva, F. H. R. Baio, L. P. R. Teodoro, C. A. da Silva Junior, R. S. Borges, and P. Teodoro, "UAV-multispectral and vegetation indices in soybean grain yield prediction based on in situ observation," *Remote Sens. Appl. Soc. Environ.*, p. 100318, 2020.

[62] J. Lu *et al.*, "Estimation of aboveground biomass of Robinia pseudoacacia forest in the Yellow River Delta based on UAV and Backpack LiDAR point clouds," *Int. J. Appl. Earth Obs. Geoinf.*, vol. 86, p. 102014, 2020.

[63] Y. Ampatzidis, V. Partel, and L. Costa, "Agroview: Cloud-based application to process, analyze and visualize UAV-collected data for

precision agriculture applications utilizing artificial intelligence," *Comput. Electron. Agric.*, vol. 174, p. 105457, 2020.

[64] "Agriculture Drones: Drone Use in Agriculture and Current Job Prospects." https://uavcoach.com/agricultural-drones/ (accessed Jan. 02, 2021).

[65] A. S. A. Ghafar, S. S. H. Hajjaj, K. R. Gsangaya, M. T. H. Sultan, M. F. Mail, and L. S. Hua, "Design and development of a robot for spraying fertilizers and pesticides for agriculture," *Mater. Today Proc.*, 2021.

[66] I. Beloev, D. Kinaneva, G. Georgiev, G. Hristov, and P. Zahariev, "Artificial intelligence-driven autonomous robot for precision agriculture," *Acta Technol. Agric.*, vol. 24, no. 1, pp. 48–54, 2021.

[67] H. Wan, Z. Fan, X. Yu, M. Kang, P. Wang, and X. Zeng, "A real-time branch detection and reconstruction mechanism for harvesting robot via convolutional neural network and image segmentation," *Comput. Electron. Agric.*, vol. 192, p. 106609, 2022.

[68] X. Yu *et al.*, "A lab-customized autonomous humanoid apple harvesting robot," *Comput. Electr. Eng.*, vol. 96, p. 107459, 2021.

[69] J. Chen, H. Qiang, J. Wu, G. Xu, and Z. Wang, "Navigation path extraction for greenhouse cucumber-picking robots using the prediction-point Hough transform," *Comput. Electron. Agric.*, vol. 180, p. 105911, 2021.

[70] R. R Shamshiri *et al.*, "Research and development in agricultural robotics: A perspective of digital farming," 2018.

[71] Y. Unpaprom, N. Dussadeeb, and R. Ramaraj, "Modern Agriculture Drones The Development of Smart Farmers 2018," *Maejo Univ.*, vol. 7, pp. 13–19, 2018.

[72] T. P. Fowdur, Y. Beeharry, V. Hurbungs, V. Bassoo, V. Ramnarain-Seetohul, and E. C. M. Lun, "Performance analysis and implementation of an adaptive real-time weather forecasting system," *Internet of Things*, vol. 3, pp. 12–33, 2018.

[73] P. H. Kulkarni and P. D. Kute, "Internet of things based system for remote monitoring of weather parameters and applications," *Int. J. Adv. Electron. Comput. Sci. ISSN*, pp. 2393–2835, 2016.

[74] A. Varghese, "Weather based information system using IoT and cloud computing," *J. Comput. Sci. Eng.*, vol. 2, no. 6, pp. 90–97, 2015.

[75] G. Chavan and B. Momin, "An integrated approach for weather forecasting over Internet of Things: A brief review," in *2017 International Conference on I-SMAC (IoT in Social, Mobile, Analytics and Cloud) (I-SMAC)*, 2017, pp. 83–88.

[76] D. K. Krishnappa, D. Irwin, E. Lyons, and M. Zink, "CloudCast: Cloud computing for short-term mobile weather forecasts," in *2012 IEEE 31st*

International Performance Computing and Communications Conference (IPCCC), 2012, pp. 61–70.

[77] M. Yan *et al.*, "Field microclimate monitoring system based on wireless sensor network," *J. Intell. Fuzzy Syst.*, vol. 35, no. 2, pp. 1325–1337, 2018.

[78] W. Qiu, L. Dong, F. Wang, and H. Yan, "Design of intelligent green-house environment monitoring system based on ZigBee and embedded technology," in *2014 IEEE international conference on consumer electronics-China*, 2014, pp. 1–3.

[79] J. P. Juul, O. Green, and R. H. Jacobsen, "Deployment of wireless sensor networks in crop storages," *Wirel. Pers. Commun.*, vol. 81, no. 4, pp. 1437–1454, 2015.

[80] K. Aliev, E. Pasero, M. M. Jawaid, S. Narejo, and A. Pulatov, "Internet of plants application for smart agriculture," *Int J Adv Comput Sci Appl*, vol. 9, no. 4, pp. 421–429, 2018.

[81] A. Goap, D. Sharma, A. K. Shukla, and C. R. Krishna, "An IoT based smart irrigation management system using Machine learning and open source technologies," *Comput. Electron. Agric.*, vol. 155, pp. 41–49, 2018.

[82] P. P. Jayaraman, A. Yavari, D. Georgakopoulos, A. Morshed, and A. Zaslavsky, "Internet of things platform for smart farming: Experiences and lessons learnt," *Sensors*, vol. 16, no. 11, p. 1884, 2016.

[83] B. M. Yahya and D. Z. Seker, "Designing weather forecasting model using computational intelligence tools," *Appl. Artif. Intell.*, vol. 33, no. 2, pp. 137–151, 2019.

[84] M. P. Raju and A. J. Laxmi, "IoT based online load forecasting using machine learning algorithms," *Procedia Comput. Sci.*, vol. 171, pp. 551–560, 2020.

[85] J. Kim, S. Cho, K. Ko, and R. R. Rao, "Short-term electric load prediction using multiple linear regression method," in *2018 IEEE International Conference on Communications, Control, and Computing Technologies for Smart Grids (SmartGridComm)*, 2018, pp. 1–6.

[86] "monographie Meknes Fes," 2009. http://www.equipement.gov.ma/Carte-Region/RegionFes/Presentation-de-la-region/Monographie/Pages/Monographie-de-la-region.aspx (accessed Apr. 26, 2020).

[87] "|." https://ma.chm-cbd.net/fes-meknes/prest_region/monographie-de-la-region-fes-meknes (accessed Aug. 21, 2021).

[88] "Place de l'agriculture dans la région de Fès-Meknès." https://www.agrimaroc.ma/fes-meknes-agriculture/ (accessed Aug. 22, 2021).

[89] "Arduino Uno Pinout, Specifications, Pin Configuration & Programming." https://components101.com/microcontrollers/arduino-uno (accessed Aug. 22, 2021).

[90] G. K. Ganjegunte, Z. Sheng, and J. A. Clark, "Evaluating the accuracy of soil water sensors for irrigation scheduling to conserve freshwater," *Appl. Water Sci.*, vol. 2, no. 2, pp. 119–125, 2012.

[91] "Mobile Apps Development - I-CLAN." https://iclan.cm/en/what-we-do/our-services/mobile-apps-development/ (accessed Aug. 22, 2021).

[92] J. D. Woodard *et al.*, "The power of agricultural data," *Science*, vol. 362, no. 6413. American Association for the Advancement of Science, pp. 410–411, Oct. 2018, doi: 10.1126/science.aav5002.

12

IoT-Based Intelligent Handicraft System Using NFC Technology

Youssef Aounzou[1], Fahd Kalloubi[2], Abdelhak Boulaalam[1]

[1]LSI Laboratory, Sidi Mohamed Ben Abdellah University, ENSA, Fez, Morocco
[2]LTI Laboratory, Chouaib Doukkali University of El JADIDA, Morocco
Email:youssef.aounzou@usmba.ac.ma; kalloubi.f@ucd.ac.ma; abdelhak.boulaalam@usmba.ac.ma

Abstract

Craft manufacturing activities are rapidly developing all over the world as the global economy grows. Meanwhile, the handcrafted industry is booming, particularly traditional crafts that have been passed down through generations. Customers, such as tourists, require more scientific and practical guidance to ensure that a craft product is genuine in these circumstances. An Internet of Things (IoT)-based craft system is presented in this chapter to track the craft product in its life cycle, especially, when it reaches its destination. The embedded system provides suggestions for end-users and the manufacturing actors using a product identification label. During the process of manufacturing products and shopping, the collected data are exchanged between the embedded sensors and the customer's smartphone. This data will then be submitted to cloud computing for processing to extract beneficial guiding information for consumers and artisans. A detailed implementation of the various system components is also presented in the rest of this chapter.

12.1 Introduction

Along with the increasing growth of the digital domain and the changing business environment characterized by large globalization, several companies are looking for new ways to value their consumers and get a competitive advantage

243

in the marketplace. Furthermore, the emergence of new technologies for companies such as the Internet of Things [1], embedded systems, intelligent products, PLM [2], and cloud integration provides rewarding opportunities for industries to improve their expertise. And as expected, recently, technology has become the major trend in a variety of fields, including health, transportation, small manufacturing, etc. in which the product equipped with auto-id technology would dominate the market. In this context, small or medium-sized enterprises (SMEs), cooperatives, and micro-enterprises do not cherish the same place in using these new technologies. The Internet of Things (IoT) technology has been classified as one of the growing information technologies that are employed to ensure the authenticity, monitoring, and traceability of products by combining them with embedded sensors.

Moreover, the labeling strategy is also adopted to ensure the authenticity of craftsmen's know-how and their heritages. It improves the production system by developing a standard for production methods to guarantee the originality, traceability, and level of quality of the artisanal product purchased.

In Morocco, the Ministry of handicrafts has established a labeling strategy for the handicraft sector to ensure the validity of those items relying on an official guaranteed mark called Morocco Handmade or National Label of Moroccan handicrafts, as detailed in [3]: The label applies to all actors in a handicraft industry: mono-craftsmen, SMEs, and cooperatives as well as the reference actors. Furthermore, it also confirms that handcrafted product meets a set of criteria prescribed by a useful guideline to guarantee a certain degree of quality.

As shown in Figure 12.1, the labeling strategy focuses on five categories of labels represented as follows: The Premium quality label guarantees the best level of quality for a product. Indeed, this label not only guarantees compliance with international standards but also the use of superior raw materials, which contribute to the production of a very high-quality product.

The Certified quality label guarantees that the product has been manufactured in compliance with international standards (EU and US in particular), thus, creating a real selling point including for export. These standards are based on objective references and have been checked reliably and impartially (external certifying bodies).

The Madmoun label is a guarantee for consumers with the respect to the quality of raw materials used, especially clay, as well as the admissible limits in terms of emissions.

The Responsible Craftsman label guarantees the respect of social and environmental standards in the manufacturing process of products. It complies with the National Charter for the Environment and Sustainable Development.

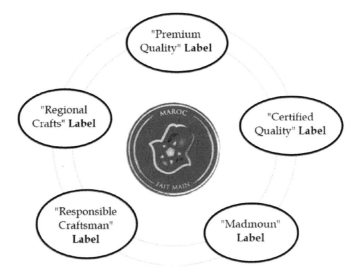

Figure 12.1 Labeling strategy of Moroccan handicraft.

The regional product label, or regional label, guarantees that a product has characteristics due to its origin. It can be a raw material or the existence of traditional know-how.

The process of attending one of these labels is based on three levels of commitments, which are each determined by different requirements that are added to encourage various actors of the craft sector to progress through a valorization of their handmade products. Since the launch of the labeling strategy by the Ministry of Crafts and Social and Solidarity Economy in 2013, 566 production units in 12 regions of Morocco were branded according to a department in charge of crafts. In the FEZ-MEKNES region, 129 production units of different types are branded with one of these labels in various fields (Wood, leather, ceramic). For more details, the following site [4] lists all labeled unit production up to now.

The problem with these labels is that they are rarely used in practice. Frequently, the manufacturing units getting these certifications do not use them in reality to develop and advertise their products. For this reason, our goal is to establish an operational embedded system in the field of handicrafts based on innovative technologies such as the Internet of Things and the labels provided by the minister of crafts in digital format, to ensure the authenticity value of handicrafts. The primary benefit of our contribution is to provide customers with digital access to product data in a fast, reliable, and secure way to protect the artistic value of these products.

The following sections of the chapter are structured as follows: The next section introduces relevant technologies and work. Section 3 describes the proposed system design, including a detailed description of several components. The system implementation is described in detail in Section 4. Finally, the last part is for the conclusion and potential future work.

12.2 Preliminary and Related Work

The concept of the Internet of Things is an extension of the basic Internet concept. It can be defined as a "logical result of the increasing technological development of several transversal fields which aims to construct a link between the physical and virtual worlds ". It does not only contain the "human" Internet but also several connected things, objects, as well as sophisticated equipment that may be used and connected to the network [5]. All physical items equipped with auto-id, mobile terminals, wired or wireless sensors, cameras, and other connecting devices fall into this category [7]. Reference [6]–[8] provides an overview of the key IoT technologies.

One of the most important aspects of the Internet of Things is the wide range of applications that are being developed every day in several areas such as fitness systems, the agriculture sector, product identification, and supply chain management.

For instance, the authors in [9] propose an IoT fitness system that collects information from several types of sensors. Furthermore, a mobile application based on NFC and AI technologies is designed to monitor the health statuses of exercisers by gathering various types of information to provide more practical guidance for exercisers. As in reference [10], it uses mobile computing development and tagged objects to create smart environments supporting human interactions. In addition, the communication between NFC technology is incorporated into mobile phones as well as smart posters distributed across the city to assist users in locating and navigating sites of interest within the city.

With the help of NFC embedded sensors and IoT technology, the paper [11] proposed a paradigm for securing art items in museum organizations. The contribution of this project was to identify the unserved need for a unified and object-specific sensory and digital data labeling system for fine art products to provide data and consciousness about the circumstances of objects and long-term competitive advantage protection through an online record of their environmental exposure.

As in the agriculture sector, the implementation of IoT technology becomes a necessity to improve efficiency, productivity, time, and cost.

Paper [12] proposes a system of agricultural services composed of NFC tags, smartphones, and field service nodes to meet farm process management needs, agricultural management human resources, and agricultural value-added services by placing NFC tags at different areas throughout the farm, such as agricultural equipment, agricultural products, in-field service nodes, and others, to provide broad management of information.

In the supply chain management sector, the Internet of Things has been used to provide real-time monitoring of supply chain employees, which employs QR codes and RFID to ensure real-time data collecting. The authors in reference [13] propose a technological model of an IoT-based tracking and tracing platform for pre-packaged food.

Furthermore, the Internet of Things is used to ensure the authenticity status of items. Research on verifying the Halal certification of food goods in Malaysia is presented [14]. In this study, an effective approach for identifying and confirming food legality is proposed by using RFID technology rather than relying only on SMS and barcodes to deliver trustworthy, thorough, and regularly updated information regarding Halal items.

In addition to the authenticity techniques mentioned in the references above used to guarantee the authenticity value of items, there are many other approaches used in different domains such as tamper-evident product packaging, traditional labeling method, and spectroscopic technique [15]. The tamper-evident packaging prevents access to the product contained in the package to maintain a basic level of security and identity. This method is widely used in the pharmaceutical field by producers. However, the disadvantage of this method is the simplicity with which it can be easily falsified by traffickers, which do not meet the needs of the consumer.

Thus, the problem with all of the aforementioned methods is the disability of merging the three criteria: consumer needs, budget, and technical constraints in a single technique. The approach proposed in this work meets this requirement by utilizing IoT technology, embedded sensors, and craft labels in digital format.

12.3 System Design

The objective is to design a smart embedded system that ensures the authenticity of Moroccan handicraft products by combining several components that are created to match these requirements and respond to the sector demand. Figure 12.2 presents a system's overall architecture that is primarily composed of the following layers:

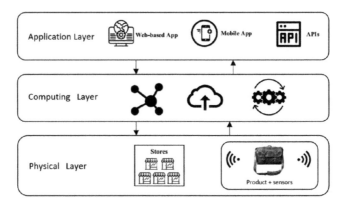

Figure 12.2   System layers.

Application layer comprises applications that are built on the services offered by the computing layer and ensures the system delivers to customers to maintain the authenticity of products.

Computing layer: The middle part of the system is used to make high computational tasks by using sophisticated processing to feed machine learning algorithms to extract useful information from collected data.

Physical Layer is a sensing layer that allows to identify and gather adequate data from diverse providers using embedded sensors that are linked with handicraft products and included in manufacturing for sending information (inputs) to the program of an embedded system. The collected data will be sent to the computing layer for analysis.

The overall architecture of an IoT-based intelligent handicraft system includes the physical layer, computing layer, and application layer. Each layer provides a set of functionalities grouped as models and processing that communicate with each other using communication technology to achieve this objective.

As shown in Figure 12.3, the physical or data acquisition layers aim to collect heterogeneous types of data from several sources like craft hand sensors, manufacturing data, explicit data, and others.

Once the data is collected from the acquisition layer, it is transferred to the server to be processed, usually, the transfer is done via standard IoT layers such as network connectivity and gateways, which are used to transmit data from the connected objects to the platform responsible for analyzing them.

At the level of the computing layer, the computational tasks are made by groups of processing models such as data processing, recommendation,

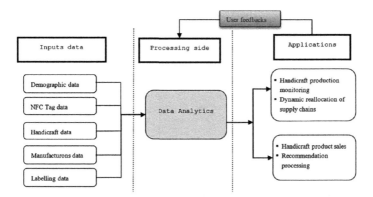

Figure 12.3 Global system schema.

and business processing to convert the collected data into useful information helping to achieve a balanced knowledge base.

The application layer consists of custom applications that make use of models established previously in the computing layer. Among these applications are many that control production, sales, and the manufacturing process of Craft Hand products.

12.3.1 Data acquisition

The data acquisition module involves the sampling and converting of physical property into digital numeric values that the cloud system can manipulate. The system can receive data from multiple sources, as shown in Figure 12.3. Each type gathers a specific kind of data. The NFC tag, that is attached to crafts, aims to assemble related data about crafts products like products-ID, version, history, etc. Then the collected data will be implicitly transmitted to the cloud system to be analyzed using various modules. When data is gathered from all the required sources, the next step is to gain insights from it using a specific protocol and tools to store, treat, transform, and analyze data to build a reliable and efficient recommendation system and provide knowledge to different contributors in the system.

In general, smartphones, are distinguished by built-in characteristics such as the capacity to detect and communicate with sensors as well as the support of communication protocols that support numerous technologies. As a result, the smartphone works as a gateway to record data, disseminate information and allow the end-user to submit comments using the HMI (Human Machine Interface) implementation.

Figure 12.4   NFC sensor used in the system.

NFC (Near Field Communication) is a type of RFID technology that is defined as a wireless short-range communication technology with a set intensity and low latency. With its ease of use, versatility, comfort, and safety, NFC will open the way for numerous creative solutions. As mentioned in reference [16], a significant factor assisting in the expansion of its application area is the development of NFC technology in conjunction with mobile phones.

In this research, NFC technology is used to identify handicraft products. As previously mentioned, this technology has many advantages that RFID technology offers in terms of cost, simplicity of use in the field such as the handicraft domain, and implicit integration into new smartphones that facilitate use.

NFC functions in three modes as detailed in reference [17]: reader/writer, card emulation, and peer-to-peer mode. Each mode has its application:

The card emulation mode allows NFC-enabled equipment to function as smart cards. It enables customers to carry out operations such as purchasing, ticketing, and controlling transportation accessibility.

Reader/Writer Mode allows users to retrieve tag information saved in the tag for future use. This sort of NFC-enabled device can also read information saved on NFC tags implanted in smart posters and displays.

Peer-to-peer mode allows two NFC-enabled devices to interact together to transfer files and data.

The proposed system uses NTG213 type NFC tags shown on the right side of Figure 12.4, which are completely user-programmable before being attached to products and may interact with customers by delivering product data as well as receiving and storing user feedback. This type of sensor also has several usability benefits including an immersive digital experience, ease of use, extreme speed, and the highest quality abs guarantee. Moreover, The NTAG 213 has been developed as standard NFC tag ICs for use in consumer applications such as retail, gaming, and consumer electronics. It is designed

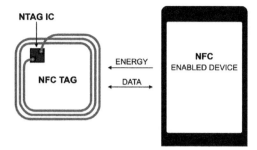

Figure 12.5 Contactless system.

to be fully compliant with ISO/IEC14443. The targeted applications of NTAG 213 include smart outdoor advertising and print, SoLoMo applications, product authentication, NFC shelf tags, and mobile companion tags. These targeted use cases include smart outdoor advertising, product authentication, mobile companion tags, bluetooth or wi-fi pairing, electronic shelf tags, and business cards. The NTAG 213 memory can also be segmented to implement multiple applications at once.

The communication in the proposed system with NTAG2213 is only possible if the IC is attached to an NFC-compatible device, as illustrated in Figure 12.5. The NTAG213 has a high-speed RF communication interface that provides data transfer at a baud rate of 106 kbit/sec when it is placed in the RF field. In these scenarios, it is the product data.

To program the NFC tags in the system, an NFC reader ACR122U, illustrated on the left side of Figure 12.4, is required. The ACR122U NFC reader is a pc-based contactless smart card reader/writer developed based on 13.56 MHz contactless technology (RFID). Compliant with the ISO/IEC18092 standard for Near Field Communication (NFC) supports not only MIFARE and ISO 14443 cards, but also the four types of NFC tags. It is perfect for secure personal identification verification. Furthermore, access control, e-payment, e-ticketing, and network authentication are some of the other uses for the ACR122U.

12.3.2 Data analysis

In general, data analysis can transform data into meaningful information that may be utilized to develop long-term understanding and make better decisions.

Extracting knowledge from collected data needs some tools and processes that help us organize the coding of data to identify relevant concepts and constructs. As shown in the reference [18], the automated data analysis can reveal information about previously undiscovered interactions between objects, their surroundings, and their users, allowing for the improvement of their behavior. Real-time data analysis integrated into physical systems, in particular, has the potential to allow novel types of remote control.

Depending on the rate of data generation, there exist two different models of data processing: Data may either be stored and analyzed as a batch, or it can be processed directly as a stream based on the speed of data generation.

Batch analysis: large amounts of data cannot be examined on a single server. It requires data distribution over multiple connected storage devices. After saving data, it may be examined in batches by distributed algorithms that perform jobs collaboratively. Each computer in a data center may have numerous cores that algorithms may leverage for parallel processing.

Several frameworks support distributed batch data analysis. Hadoop[19] is one of the most popular frameworks. It is considered a scalable data processing solution for storing and batch analyzing very large amounts of data following the map and reducing paradigm.

Analysis of streaming data: Continuously generated data from various sources (sensors, recommendations, production...) must be analyzed sequentially and gradually on a record-by-record basis. The Stream analysis will help the production unit discover new opportunities, strong and fast customer interactions, and revenue streams to increase profits data to build a reliable system.

In the proposed system, both scenarios are used, batch analysis and streaming to improve user experiences, the performance of craftsmen, and a labialization process. Batch processing is deployed when the system gets massive quantities of data from several sources, such as product data input, label data entry, and transactions that have been performed by customers. To analyze this high amount of data, we used the Hadoop MapReduce framework, which provides vast storage capacity for all types of data and analyzes this huge volume of data at once.

Stream processing is used in the system to analyze client feedback in real-time allowing data to be sent to the analysis tools as soon as it is created along with obtaining fast analysis results. There are several open-source stream processing systems available. In the suggested system, we chose Apache Kafka which seeks to provide a uniform, real-time, low-latency system for managing data streams.

As the closed-loop PLM is also adopted, the system can detect through user feedback if a manufacturer maintains the requirements of the labialization process. In addition, a recommender system is implemented to assist users in the process of choosing the products that meet their needs using the same real-time gathered data such as location and NFC tag data.

12.4 System Implementation

12.4.1 System workflow

The proposed system implementation is a sub scenario of the overall system, which is based on one type of input, sensors to ensure communication between various sides of system elements. Electronics, software, sensors, and network connectivity integrated into physical things enable data gathering and sharing. Each labeled handicraft is equipped with an NFC sensor which is customized forward by the administrator to be scanned by devices as shown in Figure 12.6. In the current context, both types of sensors are used: active and passive sensors to offer interactivity between the system components.

The content of the NFC sensors can be read automatically by the mobile application by scanning a linked tag. After parsing, a sensor starts uploading data directly to the cloud platform to verify the authenticity of the craft product.

On the other side, the server receives data collected by a smartphone app through a specially designed API software interface. When data is received, the webserver sends it to the application server, which instantly saves the raw data in a database and then processes it to create all necessary information

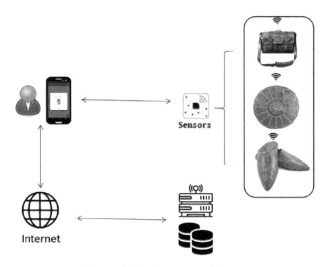

Figure 12.6   System workflow.

```
{
  // Data related to products
  "products": [
    {
      "ref_product": "001",
      "name": "Burber rugs",
      "category": "Carpet",
      "production_unit": "COOPERATIVE TAWANZA POUR TISSAGE
        TRADITIONNEL",
      "fabiration_date": "15/06/2021",
      "label_name": "Morocco Handmade",
      "detail": "traditional rug handcrafted"
    },
    {
      "ref_product": "002",
      "name": "leather bag",
      "category": "Leather",
      "production_unit": "RACHID BAJOU",
      "fabiration_date": "01/02/2020",
      "label_name": "Regional Product",
      "detail": "bag made by hand in genuine leather, natural and
        vegetable tanning, particular by the cutting of the leather
        and the interior and exterior stitching"
    }
}
```

Figure 12.7 JSON data example.

for decision making. The processed data is properly shown to consumers via dynamic layouts as shown in Figure 12.8.

To ensure a reliable and secure connection between the different architecture modules. This implementation uses REST web services based on the HTTP methods (GET, POST, PUT and DELETE) and on the JSON to format for the communication between a server and Smartphone.

12.4.2 Database design

The proposed system requires using a database management system to store a range of data.

- Craft product: includes a set of product-related attributes such as name, category, description, and other relevant information.

- Sensor tag: contains all relevant tag information as well as the product's reference to which the tag is attached.

- Production unit: stores all details about manufacturing when the products of the craft are produced such as the name, the field of activity, the labeling status, address, city, and the type of unit (company, mono-craftsman ...)

Figure 12.7 shows an example of data that is stored in the local products database. It includes relative data about crafts products and production units.

Figure 12.8 Mobile application snapshots.

12.4.3 Mobile application prototype

The prototype implemented was developed and tested by the flutter framework. It is composed of 3 parts, the first taught to the customers, the second was dedicated to the administrators and the third to the craftsmen.

To access the customer's section, no authentication is needed. The holders of devices that support NFC technology can scan the linked tag on the back of the product of the labeled craft which is already registered in the system. After scanning the NFC tag, the customer can view all relevant information about the craft, corresponding label, manufacturing details, and a proposal of recommender products product as shown in Figure 12.8.

12.5 Conclusion

This chapter has presented the architecture of an intelligent system with an early prototype based on a collection of technologies, such as the Internet of Things and cloud computing to monitor the lifecycle of handcrafted items in order the guarantee the authenticity of those types of products. The design of this proposed system, as described above, is made to ensure the well-functioning of the craft hand system through the use of several modules for collecting, analyzing, storing, processing, and transforming data to guarantee the authenticity of all handicraft own labels by transforming a craft product from the "dumb" to an intelligent item. Furthermore, the proposed system should be tested using real-world data. For that purpose, manual data collecting from the labeling process is underway to put the system into production. Related to this subject, the next investigation will be in the same direction as the implementation of a solution for transforming several labels into a virtual environment and integrating the blockchain technology into the system to facilitate real-time management.

12.6 Acknowledgments

This work was supported by the National Center for Scientific and Technical Research of Morocco identified by the following number: Alkhawarizmi/2020/28

References

[1] I. Lee, "The Internet of Things for enterprises: An ecosystem, architecture, and IoT service business model," *Internet of Things*, 2019, doi: 10.1016/j.iot.2019.100078.

[2] D. Kiritsis, "Closed-loop PLM for intelligent products in the era of the Internet of things," *CAD Computer Aided Design*, 2011, doi: 10.1016/j.cad.2010.03.002.

[3] H. Ministry, "Labeling and certification," *2021*, 2021. https://mtataes.gov.ma/fr/artisanat/qualite-et-innovation/labellisation/ (accessed Nov. 24, 2021).

[4] H. Ministry, "LE LABEL NATIONAL DE L'ARTISANAT DU MAROC," 2021. https://label.artisanat.gov.ma/consomateur?lng=fr (accessed Nov. 24, 2021).

[5] D. Gil, A. Ferrández, H. Mora-Mora, and J. Peral, "Internet of things: A review of surveys based on context aware intelligent services," *Sensors (Switzerland)*. 2016, doi: 10.3390/s16071069.

[6] L. Atzori, A. Iera, and G. Morabito, "The Internet of Things: A survey," *Computer Networks*, vol. 54, no. 15, pp. 2787–2805, 2010, doi: 10.1016/j.comnet.2010.05.010.

[7] K. Ashton, "That 'Internet of Things' Thing - RFID Journal.pdf," *RFID Journal*, 2009. .

[8] A. Whitmore, A. Agarwal, and L. Da Xu, "The Internet of Things—A survey of topics and trends," *Information Systems Frontiers*, 2015, doi: 10.1007/s10796-014-9489-2.

[9] B. Yong *et al.*, "IoT-based intelligent fitness system," *Journal of Parallel and Distributed Computing*, 2018, doi: 10.1016/j.jpdc.2017.05.006.

[10] F. Borrego-Jaraba, I. Luque Ruiz, and M. Á. Gómez-Nieto, "A NFC-based pervasive solution for city touristic surfing," *Personal and Ubiquitous Computing*, 2011, doi: 10.1007/s00779-010-0364-y.

[11] M. D. Steinberg, C. Slottved Kimbriel, and L. S. d'Hont, "Autonomous near-field communication (NFC) sensors for long-term preventive care of fine art objects," *Sensors and Actuators, A: Physical*, vol. 285, pp. 456–467, 2019, doi: 10.1016/j.sna.2018.11.045.

[12] X. Wan, T. Zheng, J. Cui, F. Zhang, Z. Ma, and Y. Yang, "Near Field Communication-based Agricultural," *Sensors*, vol. 19, p. 2, 2019.

[13] Z. Li, G. Liu, L. Liu, X. Lai, and G. Xu, "IoT-based tracking and tracing platform for prepackaged food supply chain," *Industrial Management and Data Systems*, vol. 117, no. 9, 2017, doi: 10.1108/IMDS-11-2016-0489.

[14] M. Khosravi, M. Karbasi, A. Shah, I. A. Brohi, and N. I. Ali, "An adoption of halal food recognition system using mobile Radio Frequency Identification (RFID) and Near Field Communication (NFC)," *Proceedings – 6th International Conference on Information and Communication Technology for the Muslim World, ICT4M 2016*, no. November, pp. 70–75, 2017, doi: 10.1109/ICT4M.2016.74.

[15] M. Manley and V. Baeten, *Spectroscopic Technique: Near Infrared (NIR) Spectroscopy*, 2nd ed. Elsevier Inc., 2018.

[16] B. Benyó, A. Vilmos, G. Fördos, B. Sódor, and L. Kovács, "The StoLPan view of the NFC ecosystem," *2009 Wireless Telecommunications Symposium, WTS 2009*, no. May, 2009, doi: 10.1109/WTS.2009.5068969.

[17] P. Vagdevi, D. Nagaraj, and G. V. Prasad, "Home: IOT based home automation using NFC," *Proceedings of the International Conference on IoT in Social, Mobile, Analytics and Cloud, I-SMAC 2017*, pp. 861–865, 2017, doi: 10.1109/I-SMAC.2017.8058301.

[18] R. Davies, "The Internet of Things Opportunities and Challenges," *European Parliament Briefing*, vol. 18, no. May, p. 8, 2015, [Online]. Available: http://www.europarl.europa.eu/RegData/etudes/BRIE/2015/557012/EPRS_BRI(2015)557012_EN.pdf.

[19] H. Hingave and R. Ingle, "An approach for MapReduce based log analysis using Hadoop," *2nd International Conference on Electronics and Communication Systems, ICECS 2015*, pp. 1264–1268, 2015, doi: 10.1109/ECS.2015.7124788.

SECTION 6

System on Chip and Co-design

13

SoC Power Estimation: A Short Review

Z. El Hariti[1], A. Alali[1,2], M. Sadik[1] and K. Aamali[1]

[1]NEST Research Group, Engineering Research Laboratory (LRI), ENSEM, Hassan II University of Casablanca, Morocco.
[2]Information Processing Laboratory (LTI), Faculty of Sciences Ben M'Sick. Hassan II University of Casablanca, Morocco.
zineb.elhariti@gmail.com, hakim.alali@gmail.com, m.sadik@ensem.ac.ma, aamalikaoutar@gmail.com

Abstract

In recent years, the development of embedded systems has challenged the electronics industry, driven by strong market demand for ever-evolving application functionalities. However, increasing application functionalities require an additional power budget, which consequently shortens the system's battery lifetime. Estimating an application's power consumption early in the design process creates an opportunity to extend the battery lifetime. Therefore, accurate and efficient performance analysis and estimation at all levels of abstraction throughout the design phase are becoming increasingly important. This chapter examines the concepts of single and multiprocessors, then focuses on giving a detailed description of the different existing abstraction levels. It also examines and analyses existing energy estimation techniques. It features virtual prototyping platforms that combine scalable hardware and software to estimate and evaluate energy consumption patterns.

13.1 Introduction

The number of transistors was 32 when Moore's Law [1] was predicted. However, as processing speed has increased, so has the number of transistors in processors, making them no longer scalable under Moore's law as a large number of transistors doubles every two years. Due to recent breakthroughs in silicon technology, a large number of transistors can now be housed in a

261

single chip. As a result, embedded hardware designers are increasingly using parallel and symmetric multiprocessor system-on-chip (MPSoC) designs to address the parallelism potential of demanding applications. Thanks to the advent of multicore processors, many forms of research are currently being conducted that power all modern electronic devices. In addition, the power density of advanced electronic systems is constantly increasing, resulting in high on-chip temperatures. This would damage the efficiency of the system. This can result in unreliable timing drift or even a temporary decrease in system performance. Therefore, power dissipation is a major constraint for chip architectures during device design for multiprocessor systems. A representative example is smartphones, which increase computing and communication costs due to the increasing demands of smartphone users and encourage application developers to extend their legacy applications.

Some of the most energy-hungry technologies include apps such as images, mobile games, advertising services, and social apps. When running on smartphones, the energy demands of the processors increase due to the inherent characteristics of these services. In order to use the energy balance of smartphones sensibly and to extend the life of their batteries, software optimization minimizes the interaction between the hardware components. However, software optimization methods to determine the battery-hungry part of a smartphone application require highly accurate resource-energy estimation models.

Faced with this problem, designers must predict power consumption as early as possible in the design flow to minimize development costs and time-to-market. In academic practice and current industry, power estimation utilizing low-level Computer-Aided Design (CAD) techniques is still frequently used, which is unsuitable for handling the complexities of embedded systems serving modern applications. This problem has been addressed by many systems through the development of Electronic Device Level Tools. The aim is to standardize software and hardware design and enable the introduction of virtual systems at the system level. Accordingly, computing power at the system level is seen as a critical concept to manage design complexity and achieve fast power performance.

In general, a design flow includes six levels of abstraction for system co-simulation. As indicated in Figure 13.1, these levels vary from the Algorithmic (AL) Level, which is the most abstract and imprecise, to the Layout Level, which is the most complete and realistic model. These levels are described in more detail in the following section.

The choice of level of abstraction is crucial for delivering a co-simulation prototype; the choice of a lower level of abstraction such as RTL [2] becomes

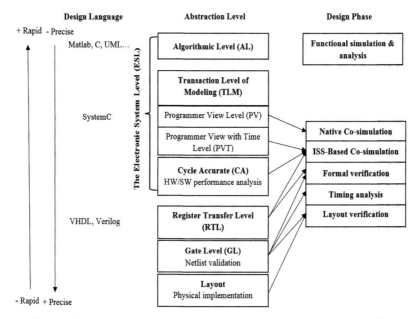

Figure 13.1 Typical levels of abstraction and design flow phases in SoCs.

a major challenge for developers and designers in contrast to the increasing complexity of the systems. This requires the use of a high level of abstraction such as TLM [3], which not only speeds up the simulation but also reduces the number of lines of code that need to be written by the developer. Co-simulation platforms, on the other hand, have to keep up with this development.

Various power estimation techniques have been proposed in the literature. SimplePower [4] and WATTCH [5] are examples of power estimation tools based on cycle-accurate simulations that provide fully accurate power estimates but require high simulation. Next, Gaspard2 [6] was proposed as a transaction-level performance estimation tool. ILPA (Instruction Level Power Analysis) was one of the examples Tiwari et al. [7] had added. This approach helps determine the processor's energy consumption by calculating the energy of each instruction to estimate the average energy consumption of a given program. This method is based on JouleTrack [8]. ILPA has been extended to FLPA (Functional Level Power Analysis) [9] for faster processing of processor power models. SoftExplorer [10] is one of the methods that include power analysis of complex and simple processors based on algorithmic and configuration parameters to create power models. However, current onboard systems are multiprocessor systems and some general studies deal with their energy estimation.

The rest of this chapter is organized as follows. Section 2 discusses the historical developments in microprocessor design, summarizes the key issues related to multi-core processors, and explains how we entered the multicore generation, as well as provides a full overview of the many levels of abstraction available. Section 3 starts by introducing the physical model. Various existing techniques used in the field of power estimation are further explained, as expressed in Section 4, and their advantages and disadvantages are examined in Section 5. Section 6 introduces a recent method for estimating performance for single and multi-core processors. Finally, further analysis possibilities were suggested in the last section.

13.2 Background

Microprocessor efficiency has improved exponentially in recent years. To achieve parallelism, techniques [11] have been split, starting with pipelining and ending with multicore processors. In this section, we shed more light on how technologies attempt to exploit any degree of parallelism by identifying some key terms as a first step. Next, we participate in the search for different levels of abstraction used.

13.2.1 Levels of parallelism

The following levels of parallelism are distinguished:

- Instruction-level parallelism: At this point, architectures use independent instructions in the instruction stream (that is, the operands of one instruction do not depend on the result of another) to execute them all simultaneously.

- Basic block level: a block is a group of instructions that all lead to a branch. With the use of sophisticated branch predictors, modern architectures use this amount of parallelism between fundamental blocks.

- Loop iterations: In superscalar architectures, for example, it is quite conceivable to perform separate iterations synchronously in these loops, since each iteration of the loop could run on an independent date.

- Tasks: A task, often referred to as a thread, is a description of a function called by a single program. To exploit this level of parallelism in multiprocessor systems, software authors must split their code into multiple threads, each running independently on a dedicated core.

13.2.2 Advances in processor microarchitecture

Over the years, as shown in Figure 13.2, several studies have been conducted to achieve better parallelism.

13.2.2.1 Single cycle processor

We have learned this method so far. The entire instruction should be executed in a single clock cycle. It's one of the simplest in terms of hardware specs that offers a practical design. As shown in Figure 13.2, all other instructions in the instruction stream must wait until all execution of an instruction is complete before proceeding. Unfortunately, this can disrupt the execution of other instructions and reduce the overall efficiency of the device.

13.2.2.2 Multi cycle processor

In single-cycle processors, each instruction takes one cycle to execute. They spend equal amounts of time following each instruction. This is one of the disadvantages of the CPU's single cycle, as it requires the computer to run at

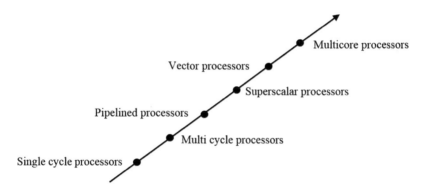

Figure 13.2   Design paradigms for microprocessors over time.

Fetch	Decode	Execute	Memory	Write Back		Single cycle processor			
					Fetch	Decode	Execute	Memory	Write Back
Fetch	Decode	Execute	Memory	Write Back		Pipelined cycle processor			
	Fetch	Decode	Execute	Memory	Write Back				

▼ Instructions

Figure 13.3   Single cycle & pipelined processor execution process.

the slowest instruction level. A multicycle processor, as the name suggests, allows multiple cycles to execute a single instruction. In this way, depending on the complexity, we can use more or fewer cycles to execute each instruction. For example, it takes three cycles to execute a branch instruction, but five cycles to execute an instruction on load.

13.2.2.3 Pipelining

CPUs are typically able to handle one or two tasks at a time. For example, most CPUs have instructions that add A to B and place the result in C. Roughly, the data for A, B, and C could be encoded directly into the statement. But it is seldom that simple when it comes to active implementation. The data is pointed to by being routed to a memory location that holds it in an address. It is seldom broadcast in raw form. It may take some time to get the data out of memory after its address has been decoded, during which time the CPU sits idle waiting for the requested data to be entered.

Most modern CPUs use a technique called instruction pipelining to reduce the time required for these steps, where the instructions go through many subunits one after the other. The single-cycle processor is divided into numerous stages in this situation; Parts of the instructions are executed simultaneously. Figure 2.2 shows how pipelining takes advantage of the parallelism of the instruction layer, which allows many instructions to run concurrently. The result is an increase in the number of instructions that can be executed in a given amount of time.

13.2.2.4 Superscalar processor

In addition to architectural design aimed at speeding up the implementation of multiple instructions, A superscalar processor is a type of microprocessor that uses instruction-level parallelism to allow multiple instructions to be executed in a single clock cycle. This refers to using multiple execution units to test these instructions. A superscalar processor is very different. In fact, there is no separation between the execution units, leading them to define the superscalar processor as the second generation of Reduced Instruction Set Computing (RISC). The concept behind RISC, however, is that computers run fast with a reduced instruction set.

13.2.2.5 Vector processor

The vector processors have a much more powerful architecture to further exploit parallelism at the data plane. By configuring a single arithmetic and logic unit, they can provide temporary reuse of a decoded instruction to execute the instruction over multiple cycles through multiple data items. Additionally, vector processors can mask the actual number of usable compute units as a micro-architectural parameter, allowing the program to determine how much work there is to do. With varying levels of parallelism, this also allows a single binary to be implemented on cores with near-optimal performance.

13.2.2.6 Multicore processors

The process is basically code that is currently being implemented. Each phase has one or more threads. When threads run concurrently, no one has to wait for the other. Multithreading wasn't even used in the traditional uniprocessor. The uniprocessor gives the impression that all threads are running at the same time, but in reality, the same process threads are switched quickly. At the thread level, today's multicore processors make effective use of parallelism. They seemed to extract as much parallelism as possible from the threat level of a multicore processor. As a result, the overall performance of the system is changed significantly. Various difficulties also arise, such as B. building the required memory structure, designing interconnection networks, and ensuring processor reliability and validity.

13.2.3 Abstraction levels classification

13.2.3.1 Layout

When the location and design of each transistor are carefully checked and precisely known, then we are talking about the most precise description of a chip. The data is then converted to an industry-standard format, typically Graphic Design System (GDSII), and sent to a semiconductor foundry after verification is complete. The data patterns representing the transistors and their connections are finally converted into masks by the foundry. Electronic Design Automation (EDA) tools such as place-and-routing tools or schematic-driven layout tools are used to create a modern Integrated Circuit (IC) layout using IC layout editing software.

13.2.3.2 Gate level

The gate-level abstracts a lot of information by simply focusing on logic gates (AND, OR, flip-flops, etc.) and their connections. This description is then loaded into a place-and-route EDA tool as an output netlist from the RTL synthesis tool, typically represented using Verilog gate-level primitives.

13.2.3.3 Register transfer level

RTL specifies a physical system implementation using registers and a data flow description of the transfers between them. Each wire will be described later, but its specific value will only be determined with each cycle. At this level, models are written in hardware description languages (HDLs) such as VHDL, SystemVerilog, or Verilog.

EDA synthesis tools perform the translation to gate level, allowing for automatic circuit optimization in terms of surface, power, and timing. Several industrial tools, such as

Philips' Petrol [12], allow RTL level consumption evaluation. To acquire exact values for consumption elements, designers typically use simulation with input triggers, which minimize estimation error. Compared to the findings produced using SPICE [13-14] this error can range from 10% to 15%.

13.2.3.4 Cycle accurate level

At the Cycle-Accurate level, the model is precisely represented in terms of execution time. What happens at each clock cycle's interval affects the behavior of a CA component. It has the same wires as RTL and has the same value at each clock cycle. The designer is free to design the component's internals. Using a standard programming language, a CA internal behavior generally implements and computes the various outputs based on the current and previous inputs (such as C or SystemC). A description of the processor's internal micro-architecture (pipeline, branch prediction, cache, etc.) is accomplished at the processing level. However, a precise bit-accurate communication protocol is performed at the communication level.

13.2.3.5 Transactional level modeling

TLM is a high-level modeling approach describing a Virtual Prototype (VP) of a hardware design portion using high-level description languages such as SystemC [15].

In the literature [16-17-18-19-20], there have been several definitions and classifications of different TLM levels. These indicate a lack of agreement on the concept of a TL model. Nonetheless, there are several things that all of these approaches have in common. Firstly, in the TLM environment, a transaction is defined as the synchronization or exchange of data structures and/or control information between two components [21-22]. The use of specific communication techniques known as through channels [23] simplifies the passing of transactions. Secondly, the concepts of communication and computation are separated.

Therefore, TLM is provided as several sub-levels classification. We use the Programmer View (PV) and Programmer View with Time (PVT) levels to classify TLM levels, as shown in Figure 13.1. These levels correspond to our interested type models. The system architecture is considered at these two sub-levels.

13.2.3.6 Algorithmic level

At this stage, the application is algorithmically described using a standard specification or existing documentation. At this level, models are described with a higher description language such as Matlab, Python, C, or C++. They are then functionally examined and validated in order to most effectively separate the application into hardware and software tasks. Model-based engineering approaches (e.g. Mealy and Moore machines) and languages (e.g. Unified Modeling Language (UML)[24] and Architecture Analysis and Design Language (AADL)[25]) are commonly used to specify and analyzing an application algorithmically.

13.3 Physical Power Models

In general, the system draws power during the activities involved. This power could be expressed in three sections: leakage power, dynamic power, and short circuit power as shown in Equation 13.1. If this power is constant, the energy is simply multiplication by the duration. The n is used here to indicate whether the system is currently drawing dynamic power and to represent the state of the system. More specifically, n=0 when the processor is in the idle state and n=1 when the processor is in the active state.

$$P_T = nP_{dynamic} + P_{short\text{-}circuit} + P_{leak} \qquad (13.1)$$

$$E_T = P_T T_{execution} \qquad (13.2)$$

13.3.1 Leakage power

Leakage power consumption mainly results from the leakage current [26], and is expressed as

$$P_{leak} = N_{gate} V_{dd} I_{leak} \qquad (13.3)$$

Where N_{gate} is the number of gates, V_{dd} is the supply voltage, and I_{leak} is the leakage current. The I_{leak} is described by a nonlinear exponential equation.

$$I_{leak} = I_s \left(AT^2 e^{(v_1 v_{dd} + v_2)/T} + Be^{(v_3 v_{dd} + v_4)} \right) \qquad (13.4)$$

Where I_s is the leakage current at a particular reference temperature and supply voltage, T is the operating temperature, v_1, v_2, v_3, v_4, A, and B are empirically determined, technology-dependent constants. Equation 13.4 demonstrates the existing relationship between the leakage power and temperature. However, the high-order and nonlinear terms make (13.4) excessive to perform real-time feasibility analysis. As reported in [27], the leakage current changes super linearly with the temperature. So, using a linear model approximation, the leakage-temperature dependence can significantly simplify the leakage model while maintaining acceptable accuracy. Therefore, as in [28], we model the leakage power of the processor P_{leak} at the supply voltage/speed $(v_k; s_k)$ as

$$P_{leak}(k) = (\alpha_k + \beta_k T)v_k \qquad (13.5)$$

Where α_k and β_k are constants depending on processor P_k.

13.3.2 Dynamic power

To build the functionality of each gate through logical gates implemented in CMOS chips, two complementary types of transistors are used, such as PMOS and NMOS. The VGND ground voltage is connected to one NMOS transistor terminal, while one PMOS transistor terminal is typically linked to the VDD voltage supply. The schematics of an inverter gate consisting of one transistor for PMOS and one transistor for NMOS are presented in Figure 13.3. To explain how the gate works, we'll start with a logic 0 input voltage. In this scenario, the NMOS transistor has a very high off-state resistance (ideally ∞), but the PMOS transistor has a very low on-state resistance (preferably 0), thus there is a channel to charge the load capacitance CL until the output voltage hits VDD. The load capacitance, CL, is the total

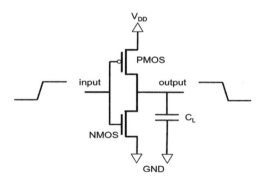

Figure 13.4   CMOS inverter.

capacitance resulting from the two transistors' output diffusion capacitances, fan-out gate input capacitances, wiring capacitance, and parasitic capacitance. As the input voltage approaches logic 1, the NMOS transistor is in an on state with very low resistance (ideally 0), the transistor of PMOS is in an off state with extremely high resistance, and there is a direction to discharge the loads to the ground until the output voltage is 0.

The energy consumed during each per cycle, is equivalent to $C_L V_{DD}^2$. Therefore, the dynamic power consumed per gate $P_{dynamic\_gate}$ referring to the switching energy per second, is governed by the following equation.

$$P_{dynamic\_gate} = SC_L V_{DD}^2 f \qquad (13.6)$$

S represents the activity factor, which indicates the number of switching cycles per second. If a circuit has N gates, then the total dynamic power is equivalent to Equation 3.7.

$$P_{dynamic} = \sum_i^N S_i C_{L_i} V_{DD}^2 f \qquad (13.7)$$

Where S_i and C_{L_i} are the switching activity and load capacitance of gate i respectively.

Again, we can control the amount of complex power consumed simply by changing frequency-voltage pairs. This method was used to get a different ratio of energy per command. Introduced commercially as Intel's SpeedStep technology [29] and AMD's PowerNow [30].

13.3.3 Short-circuit power

The short circuit power is another power factor involved in switching. There is a brief time when both the NMOS and PMOS transistors are on and current flows from the supply terminal to the field when the input signal does not have a strong transition edge. And during a switching process, the short-circuit power is consumed and is therefore proportional to the dynamic power consumption. The exact value of the input and output signals is calculated by the slopes or transition times. With a suitable circuit design, the short-circuit power is typically around 10% of the dynamic power [31]. Therefore, it is ignored in most approaches. Conceptually, generic power models depend on various parameters such as: frequency and voltage as shown in Equations 13.3 and 13.6, and statics such as capacitance, leakage current, temperature, and other physical values depending on the technology.

13.4 Power Estimation Techniques

For large circuits, the low-level tools demand a significant simulation time, making them unusable for complex MPSoC. However, these tools are still impractical to implement in the new design flow because they require knowledge of the circuit details, even if they offer good accuracy. In this way, approaches relating to energy efficiency can be evaluated. This section provides a high-level overview of the latest different power modeling approaches at different levels of abstraction. The power consumption models would be content with high-level models dealing with functional units of programs, instructions and with no electrical understanding of architecture [32].

Many solutions have chosen to instrument the code when the program is run on a given platform to get an early estimate of the power consumed. These solutions are incomplete because for a reasonable estimate of consumption we should consider both software and hardware. Another concept is to study the energetic characteristics of the circuits during the virtual prototyping phase, which allows for architectural experiments based not only on power requirements but also on the power consumed. First of all, consumption estimates during the virtual prototyping process consist in determining the events whose occurrence leads to certain energy consumption. It is then a matter of assigning energy to each of these events; we are talking about characterization. Finally, validation of these consumption models should show that they can determine how much energy a device consumes while modifying the software application. We will introduce some of these solutions in this section. The finesse of the models offered and the characterization of the different models would determine the difference between the solutions available.

13.4.1 WATTCH

WATTCH [5] is one of the first scientific tools for system-level consumption estimation. In 2000, the first article describing this tool, created at Princeton University in collaboration with Intel, was published. WATTCH allows analysis and optimization of microprocessor performance with an acceleration factor of 1000 compared to tools at the mask drawing level and an error of less than 10%. This tool provides models of configurable consumption for the components of a microprocessor. Equation 13.8 shows the total power consumed for the various sub-blocks of a multiprocessor system.

$$E_T = \sum_i E_T(i) \qquad (13.8)$$

Where i defines the type of component.

The energy in (13.9) consumed by each subblock is equal to the product of the occurrence number of an event $N(i)$ by the elementary energy consumed at each occurrence of this event.

$$E_T(i) = N(i, j) \, e(i, j) \qquad (13.9)$$

Where i defines the type of component and j defines the type of events.

13.4.2 AVALACHE

One of the most common tools based on virtual prototyping is the AVALANCHE tool [33], which was the subject of an NEC laboratory publication in 2002. AVALANCHE offers consumption models for the following components: cache, memory, and processor.

Equation 13.10 presents the processor model defined by two types of events:

$$E_{proc} = E_0 N_0 + E_1 N_1 \qquad (13.10)$$

N_0 and N_1 are respectively the number of cycles when the processor is idle and the number of cycles during which the processor is active. They are obtained from an ISS simulator. E_0 and E_1 represent the elementary energies.

13.4.3 PowerVIP

PowerViP [34] is an experimental tool designed to estimate consumption, implemented in a TLM-level simulator, [35] ViP. This tool was created in collaboration with Yonsei University in Korea by Samsung Electronics. As for AVALANCHE, the processor is modelled by two states busy a nd idle.

13.4.4 Hybrid System Level power consumption estimation (HSL)

One of the academic methods for hybrid consumption estimation for systems-on-chip is HSL [36]. It was published in 2011 as an established collaboration between the University of Lille, the University of Valenciennes, and the University of Southern Brittany. The proposed consumption models are based on the FLPA method (Functional Level Power Analysis) [37].

With HSL, characterization is achieved by performing multiple post synthesis physical measurements on the FPGA. Various applications are deployed on top of the architecture, which in principle allows each block in the system to be simulated independently. Then the total consumption of the system is measured for each software application. Finally, the elementary energies of the model are obtained by regression. The regression consists of modeling, formally, the relationship between the total energy E_T and the frequency of occurrence of the different types of events N_i. In general, the linear model can be written as follows:

$$E_T(i) = \sum_j N(i,j)e(i,j) \qquad (13.11)$$

13.4.5 Early Design Power Estimation (EDPE)

One of the main approaches is based on consumption models per component in order to derive the energy dissipated by the overall system from the description of CABA (Cycle Accurate Byte Accurate) or TLM (Transaction Level Model) level. The models proposed by EDPE [38] were integrated into the virtual prototyping library SoClib. In this way, during the architectural exploration phase, the designer not only has the temporal and spatial properties of his circuit but also an accurate estimate of his energy consumption.

In much work, the processor has been modeled by as many events as the types of instructions it can process. This model has two main disadvantages, on the one hand, its strong dependence on the training game of the processor used, and on the other hand the complexity of the model in terms of the number of events. For these two reasons, the authors considered a simple model based on three types of events: idling, running, and stuttering.

$$E_{proc} = E_{IDLE}\, N_{IDLE} + E_{RUNNING}\, N_{RUNNING} + E_{STUTTERING}\, N_{STUTTERING} \quad (13.12)$$

13.4.6 Recent power and temperature modelling method

Claude Helmstetter et al. [39] presented an approach that expresses the instruction power in terms of component states (i.e., off, idle, etc.). This task could be realized by function calls in the SystemC code that represents the model of the hardware. The proposed solution was applied to a developed system. The first part represents the SystemC/TLM model, which communicates with an Analytical Model of Temperature in Microprocessors (ATMI) [39] via the ATMI wrapper. For best control, a graphical user interface (GUI) is used to monitor and control the simulation. During the simulation and based on the power consumption as the power density of each area, ATMI calculates the temperature at a regular pace, such as once every millisecond. The ATMI is called directly from SystemC code since it is packaged as a C library. Calculating the temperature of the components and the given power densities is simply a function call to the ATMI library. As an example shown, when the temperature reaches some thresholds, the temperature sensor defines a call-back method that triggers an interrupt.

The general formula is expressed by Equation 13.13.

$$P = P_{static} + P_{dynamic} = VK_1(1 + K_2T) + FV^2\alpha K_3 \qquad (13.13)$$

13.5 Discussion

When performing tasks on on-chip multiprocessor systems, the new trend calls for optimizing legacy applications for efficient use of battery resources. Therefore, running its applications helps developers reconsider their device design for efficient use of battery resources in the early stages of development. In general, a power estimation approach relies on hardware-level or higher-level estimation to predict consumption during the execution of a task. On the one hand, low-level estimates were the first work targeting the level of abstraction that provided the most data. On the other hand, the demand for high-level power estimation simulators has increased recently, which enables early exploration of the design field. WATTCH provides a reasonable balance between simulation precision and speed; however, not all parts that make up the system in particular connections are modeled.

WATTCH provides a reasonable balance between simulation precision and speed; however, not all parts that make up the system in particular connections are modeled. The characterization approach is not homogeneous. Depending on the design of each node, calculating the power consumption at each connection to a sub-block uses four different models, making it difficult to generalize. Thus, in addition to the missing modeling of the circuits, the classification is also the weak point of WATTCH. AVALANCHE offers a reasonable balance between accuracy and speed when it comes to estimating the energy consumed by the processor alone; however, the accuracy is slightly degraded when considering a complete system. In fact, the power consumption of the system measured with AVALANCHE does not take into account the interconnection consumption. The models of the various components are delimited from each other in different ways, which makes it difficult to characterize them in general.

For the sake of accuracy, the designers of the Power VIP tools decided to use a different characterization method for each part of the studied platform: memory sheets and RTL level estimation tool for the rest. When developing the bus model, PowerViP uses a different architecture under different conditions. We recognize that this heterogeneity increases the time to build useful models for hardware components other than the current ones.

In HSL, the basic principle is to use the regression method to classify specific energies. The benefit of this approach is fairly general. However, for regression characterization to be effective, it is necessary to write test vectors (i.e., different applications), each capable of activating a component. These test vectors were not provided by the authors. However, we believe that the operation of several components, especially the processor and its cache, is difficult to separate. This role is much more difficult to fulfil when the multicore model is involved. Other research efforts propose an EDPE technique that could be used to choose between different implementations of the same software application on the same hardware architecture. Although EDPE is used to compare the power consumption for different hardware platforms or to compare different software algorithms used on the same hardware platform, it is often only used for homogeneous architectures. The power and temperature estimates can be captured early by using recent power and temperature modeling methods, but still lack accuracy as long as the static power model is based on a linear equation. The framework used improved simulation speed, with some TLM modules using a DMI (Direct Memory Interface) technique. It speeds up memory access by providing the initiator with a point to the memory array. Therefore, the initiator uses the memory pointer directly when accessing memory, rather than creating a transaction that travels through the bus.

13.6 Proposed Technique

Based on the discussion above and by reference to the comparison in Table 6.1, we were able to propose a technique that we believe offers the best balance between estimating power and temperature and overall performance. The technology will use a LIBTLMPWT-based hardware/software co-simulation [40] with the implementation of generic system-level performance models, in particular the static performance model. Accuracy is improved in this way. As the first promising work, reported in [41], the technique was applied to a single soft processor with shared memory, one for data and the other for instructions. The results show the platform performance through the different usages of the Game of Life application for 65nm and 22nm technology nodes. The technique could be extended to different architectures that can

Table 13.1 Code analysis-energy estimation and evaluation of modeling schemes.

Related literature	Abstraction level	Profiled instruction type	Target processor	Claimed accuracy, %	Benchmark(s)
5	Architectural level	Scenarios as a c program	Intel Pentium Pro and Alpha 21264	90	SPECint95, SPECfp95
43	System level	Application	ARM926EJ-S	95	Dhrystone
44	Micro-architectural level	Pthread APIs	Niagara Niagara 2 Alpha 21364 Xeon Tulsa	N/A	PARSEC[45], SPLASH-2[46], SPEC CPU 2006[47]
38	System level (TLM or CABA)	Application	MIPS32	93–99	Merge Sort, Insertion Sort , Quick Sort, Filter
39	System level (TLM)	Application	Microblaze	N/A	Game of life[48]

include more than one processor, which describes its flexibility [42]. Our methodology is currently being further explored and our future work is to validate it using hardware simulations.

13.7 Conclusion

In this chapter, we provide a comprehensive literature review on performance estimation techniques for single and multi-processors by presenting various techniques used and examining their problems, advantages, and limitations. An analysis was performed for each technique. Finally, we have proposed a new methodology with which the results obtained are comparable and sometimes better compared to others. It is very clear from the discussion that a system-on-chip design does not have a perfect way of estimating the power. It depends not only on the architecture or microarchitecture to have the right accuracy in the estimation but also on the proposed model and the way it will be implemented, without forgetting the specifics of the language used. The combination of certain proposed techniques makes it possible to provide an optimal solution. Future work proposes an assessment of a large platform containing multiple IPs and heterogeneous multiprocessors.

References

[1] Shalf J. 2020. The future of computing beyond Moore's Law. Phil. Trans. R. Soc. A 378: 20190061. http://dx.doi.org/10.1098/rsta.2019.0061.

[2} A.Raghunathan , S.Dey, Niraj K. Jha, 1996. Register- Transfer Level Estimation Techniques for Switching Activity and Power Consumption. In IEEE

[3] Donlin, "Transaction level modeling: Flows and use models". IEEE/ACM/IFIP International Conference on Hardware/Software Codesign and System Synthesis, ACM, pp. 75–80, 2004

[4] W. Ye, N. Vijaykrishnan, M. Kandemiret al. "The design and use of simplepower: A cycle-accurate energy estimation tool". Proceedings of the 37th Annual Design Automation Conference, ACM, pp. 340–345, 2000.

[5] D. Brooks, V. Tiwari, and M. Martonosi. "Wattch: a framework for architectural-level power analysis and optimizations". ACM SIGARCH Computer Architecture News, 28(2):83–94, 2000.

[6] R.B. Atitallah, S. Niar, J.-L. Dekeyser, "Mpsoc power estimation framework at transaction level modeling". The 19th International Conference on Microelectronics, ICM, IEEE, pp. 245–248, 2007.

[7] V. Tiwari, S. Malik, A. Wolfe, "Power analysis of embedded software: a first step towards software power minimization". IEEE Trans. VLSI Syst. 2 (4), pp. 437–445.

[8] Sinha, A.P. Chandrakasan, "Jouletrack: A web based tool for software energy profiling". Proceedings of the 38th Annual Design Automation Conference, ACM, pp. 220–225, 2001.

[9] J. Laurent, N. Julien, E. Martin, "Functional level power analysis: An efficient approach for modeling the power consumption of complex processors". Design, Automation and Test in Europe Conference, IEEE Computer Society, pp. 666–667, 2004.

[10] E. Senn, J. Laurent, N. Julien, E. Martin, "Softexplorer: estimation, characterization, and optimization of the power and energy consumption at the algorithmic level". Integrated Circuit and System Design. Power and Timing Modeling, Optimization and Simulation, Springer, pp. 342–351, 2004.

[11] Sinha, A.P. Chandrakasan, "Jouletrack: A web based tool for software energy profiling". Proceedings of the 38th Annual Design Automation Conference, ACM, pp. 220–225, 2001.

[12] Rafael Peset Llopis and Kees Goossens. « The petrol approach to high-level power estimation ». In: Proceedings. 1998 International Symposium on Low Power Electronics and Design (IEEE Cat. No. 98TH8379). IEEE. 1998, pp. 130–132.

[13] url: https://www.xilinx.com/products/design-tools/ise-design-suite.html.

[14] Andrew Adamatzky. Game of life cellular automata. Vol. 1. Springer, 2010.

[15] SystemC modeling language. url: http://www.systemc.org

[16] Reinaldo A Bergamaschi and Yunjian W Jiang. « State-based power analysis for systems-on-chip ». In: Proceedings 2003. Design Automation Conference (IEEE Cat. No. 03CH37451). IEEE. 2003, pp. 638–641.

[17] Suraj Singh Bhadouria, Nikhil Saxena, and PG Scolar. « An Ultra Low Power 5–Phase Ring Oscillator using Lector Technique ». In: ().

[18] Dominique Blouin and Eric Senn. « Cat: An extensible system-level power consumption analysis toolbox for model-driven design ». In: Proceedings of the 8th IEEE International NEWCAS Conference 2010. IEEE. 2010, pp. 33–36.

[19] Andrea Bona, Vittorio Zaccaria, and Roberto Zafalon. « System level power modeling and simulation of high-end industrial network-on-chip ». In: Ultra low-power electronics and design. Springer, 2004, pp. 233–254.

[20] Tayeb Bouhadiba. « 42, A Component-Based Approach to Virtual Prototyping of Heterogeneous Embedded Systems ». PhD thesis. Institut National Polytechnique de Grenoble-INPG, 2010.

[21] Tayeb Bouhadiba et al. « Co-simulation of functional systemc tlm models with power/thermal solvers ». In: 2013 IEEE International Symposium on Parallel & Distributed Processing, Workshops and Phd Forum. IEEE. 2013, pp. 2176–2181.

[22] Mark Burton et al. « Transaction Level Modelling: A reflection on what TLM is and how TLMs may be classified. » In: FDL. Citeseer. 2007, pp. 92–97.

[23] Adam Rose et al. « Transaction level modeling in SystemC ». In: Open SystemC Initiative 1.1.297 (2005).

[24] OMG Unified Modeling Language. url: http://www.uml.org.

[25] Peter H Feiler et al. « An overview of the SAE architecture analysis & design language

[26] Y. Liu, R. Dick, L. Shang, and H. Yang, "Accurate temperature dependent integrated circuit leakage power estimation is easy," in Proc.Int. Conf. Design Automation and Test in Europe, pp. 1526-1531, 2007.

[27] G. Quan and V. Chaturvedi, "Feasibility analysis for temperature constraint hard real-time periodic tasks," IEEE Trans. Industrial Informatics, vol. 6, no. 3, pp. 329-339, 2010.

[28] S.M. Jagtap, V.J. Gond, "Study the Performance Parameters of Novel Scale FINFET Device in nm Region", 347 *the first International Conference on Electronics, Communication and Aerospace Technology (ICECA)*, 348 Tamilnadu , India, 2017.

[29] Nikhil Bansal et al., "Speed Scaling for Weighted Flow Time", SIAM Journal on Computing, Volume 39, Issue 4, 2009.

[30] Y. Liu, R. Dick, L. Shang, and H. Yang, "Accurate temperature dependent integrated circuit leakage power estimation is easy," in Proc.Int. Conf. Design Automation and Test in Europe, pp. 1526-1531, 2007.

[31] J. Zhou, T. Wei, M. Chen, J. Yan, X. Sharon Hu and Y. Ma, "Thermal-Aware Task Scheduling for Energy Minimization in Heterogeneous Real-Time MPSoC Systems" in IEEE Transaction on Computer-Aided Design of Integrated Circuits and Systems, vol. 35, pp. 1269-1282, 2016.

[32] M. Casas-Sanchez C. J. Bleakley and J. Rizo-Morent, "Software level power consumption models and power saving techniques for embedded dsp processors". Journal of Low Power Electronics, 2:281290, 2006.

[33] J. Henkel and Y. Li. "Avalanche: an environment for design space exploration and optimization of low-power embedded systems". Very Large Scale Integration (VLSI) Systems, IEEE Transactions on, 10(4):454–468, 2002.

[34] I. Lee, H. Kim, P. Yang et al. "PowerViP: Soc power estimation framework at transaction level". In Proceedings of the 2006 Asia and South Pacific Design Automation Conference, pp.551–558. IEEE Press, 2006.

[35] Soo-Kwan Eo, "Vip: A practical approach to platform-based system modeling methodology". jounal of semiconductor technology and science, 5(2):89–101, 2005.

[36] S. K. Rethinagiri, R. Ben Atitallah, S. Niar et al. "Hybrid system level power consumption estimation for fpga-based mpsoc". In Computer Design (ICCD), 2011 IEEE 29th International Conference on, pp. 239–246. IEEE, 2011.

[37] J. Laurent, N. Julien, E. Martin, "Functional level power analysis: An efficient approach for modeling the power consumption of complex processors". Design, Automation and Test in Europe Conference, IEEE Computer Society, pp. 666–667, 2004

[38] K. Z. Elabidine, "EDPE: Early Design Power Estimation", thesis, University of Pierre et Marie Curie – Paris, France, 2014.

[39] P. Michaud and Y. Sazeides. "ATMI: analytical model of temperature in microprocessors". Third Annual Workshop on Modeling, Benchmarking and Simulation (MoBS), 2007.

[40] M. Moy, C. Helmstetter, T. Bouhadiba et al. "LIBTLMPWT: Modeling Power Consumption and Temperature in TLM Models". Leibniz Transactions on Embedded Systems (LITES), vol. 3, n. 1, pp. 03:1-03 , June 2016.

[41] Z. El Hariti, A. Alali, M. Sadik et al. "Co-Simulation of Power and Temperature Models at the SystemC/TLM for a Soft-core Processor". Advances in Materials Science and Engineering, vol. 2020.

[42] A. J. Brandenburg, B. Stabernack, "Simulation-based HW/SW Co-Exploration of the Concurrent Execution of HEVC Intra Encoding Algorithms for Heterogeneous Multi-Core Architectures", Journal of Systems Architecture, vol. 77, 2017.

[43] V. Tiwari. Logic and system design for low power consumption. Princeton University, 1996

[44] Jeff Janzen. Calculating memory system power for SDDR SDRAM. Designline, 10(2), 2001.

[45] C. Bienia, S. Kumar, J. P. Singh, and K. Li, "The PARSEC Benchmark Suite: Characterization and Architectural Implications," in *PACT*, 2008.

[46] S. C. Woo, M. Ohara, E. Torrie, J. P. Singh, and A. Gupta, "The SPLASH-2 programs: Characterization and Methodological Considerations," in *ISCA*, 1995.

[47] J. L. Henning, "Performance Counters and Development of SPEC CPU2006," *Computer Architecture News*, vol. 35, no. 1, 2007.

[48] A. Adamatzky, Game of Life Cellular Automata, Fac. of Computing, Engineering and, Mathematical Sciences (CEMS) University of the West of England Bristol United Kingdom, Springer, London, 2010.

14

Hardware/Software Partitioning Algorithms: A Literature Review and New Perspectives

A. Iguider, K. Bousselam, O. Elissati, M. Chami, and A. En-Nouaary

Institut National des Postes et Telecommunications, Lab. STRS, Av. Allal El Fassi, Madinat Al Irfane, Rabat, Morocco.
Email adil.iguider@gmail.com, {bousselam, elissati, chami, abdeslam}@inpt.ac.ma

Abstract

The design complexity of embedded systems is constantly increasing. For this reason, robust approaches must be adopted. The principal aims behind adopting such approaches are: reducing the development time, achieving the high possible performance, and meeting the functional specifications of the system. The compound design (codesign) is among the most powerful methodologies used to fulfill the aforementioned requirements. The Hardware Software Partitioning (HSP) is one of the most important steps in the codesign. Its role is to define the best possible partitions for the hardware and software functionalities. This chapter focuses on major algorithms proposed in the domain of hardware-software partitioning. It sheds light on some traditional exact algorithms and several heuristics and meta-heuristic algorithms. New perspectives are also presented including proposed solutions based on some powerful techniques such as the Lagrangian relaxation method, game theory, and Balas method. The main common objective accentuated the optimization of the involved parameters in the HSP process as well as the speedup of the algorithms.

14.1 Introduction

An Embedded System (ES) is generally composed of several hardware (HW) blocks and software (SW) tasks implemented together in one chip. Depending

283

on the complexity of the system, the SW blocks are running either on a single processor or on multiprocessors. Other than meeting its functional specifications, and ES faces several challenges related to non-functional requirements such as the marketing time, the execution time, the hardware cost, power consumption, and so on. To achieve these requirements, new approaches have to be adopted while designing an ES. One of the distinguishable approaches is the codesign methodology. The codesign is composed of a set of engineering processes grouped into four processes: the Co-specification process, Co-synthesis process, Co-simulation process, and Co-verification process. The Co-specification describes the functionalities of the system whereas the Co-synthesis defines the architecture of the system. The Co-simulation step, on the other hand, simulates the hardware and the software simultaneously before the prototyping. Lastly, the Co-verification phase mathematically verifies whether the specifications of the system are met. HSP is part of the Co-synthesis. It is considered to be a vital process in the codesign. The objective of the HSP is to split the system's functionalities into two major sets (hardware set H and software set S), and choose for each functionality the best implementation possible (HW or SW).

Many studies have been conducted in the literature considering the following parameters: the execution time (performance) and the hardware area (cost). For instance, in [1], a proposed solution was based on the Genetic Algorithm. While in [2], the solution was based on the Tabu Search Algorithm. In [3], the Simulated Annealing algorithm was used to solve the HSP problem. Other studies also considered power consumption which is especially important for portable devices. For example, in [4] (Hierarchical Clustering), [5] and [6] the proposed algorithms dealt with the problem on the basis of the three parameters: execution time, hardware cost, and power consumption. Other studies added more parameters by considering, for example, the access to the shared memory and the IP (Intellectual Property) reuse. There are mainly two families of algorithms: the family of classical algorithms and the family of modern algorithms. This chapter gives the principles of some of the significant algorithms of each family and the advantages and disadvantages of each of them.

The perspectives present new heuristic algorithms. These perspectives can be grouped into three categories: The first category is based on the Lagrangian relaxation method. The second category is based on the GO game, while the third category is built upon the Balas method and 0-1 Knapsack algorithm.

The rest of the chapter is organized as follows: An overview of the partitioning problem is explained in Section 14.2. Section 14.3 reviews the most exact algorithms used to solve the problem. The solutions based on classical heuristic algorithms are presented in Section 14.4. Modern proposed approaches are explained in Section 14.5. The new perspectives are presented in Section 14.6. Finally, Section VII is dedicated to conclusions and possible future research.

14.2 Overview of Partitioning Problem

The general Hardware/Software Partitioning problem consists of assigning blocks $B = \{B_1, B_2, B_3, \ldots, B_n\}$ to m partitions $P = \{P_1, P_2, P_3, \ldots, P_m\}$, such that:

- $P_1 \cup P_2 \cup \ldots \cup P_m = B$
- $P_i \cap P_j = \emptyset, \forall i, j: i \neq j$
- Minimize the objective function $f(P)$ subject to constraints functions

Each block B_i represents the functionality of the system and each partition P_j represents the target architecture. A general view of the partitioning problem is given in Figure 14.1.

As described in [7], there are two target architectures: the software architecture which includes DSP and ASIP, and the hardware architecture which includes FPGA and ASIC. The objective function $f(P)$ evaluates global partitioning and expresses the factors to minimize the execution time, hardware cost, power consumption, etc. The yielded partition must optimize the involved factors and meet the constraints.

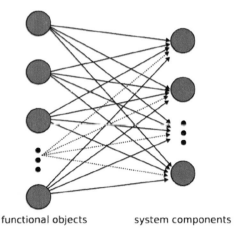

functional objects   system components

Figure 14.1   Schematic view of the partitioning problem.

14.3 Exact Algorithms

Exact Algorithms were among the first solutions proposed to the HSP problem. The most used algorithms are: Integer Linear Programming (ILP) method, Dynamic Programming (DP) method, and Branch and Bound (BB).

14.3.1 Integer linear programming

Linear Program (LP) formulation is composed of the following elements:

- Binary variables for HSP problem

- The constraints functions

- The objective function

As an example, an HSP problem can be formulated as in 14.1 using ILP formulations.

$$x_{i,k} \in \{0,1\}, 1 \le i \le n, 1 \le k \le m$$

$$\sum_{k=1}^{m} x_{i,k} = 1, \ 1 \le i \le n$$

$$minimize \sum_{k=1}^{m}\sum_{i=1}^{n} x_{i,k} * c_{i,k}$$

$$subject \ to : \sum_{i=1}^{n} x_{i,k} \le h_k, 1 \le k \le m \qquad (14.1)$$

The binary variable $x_{i,k}$ indicates whether the block belongs to the partition P_k. $c_{i,k}$ is the cost of the block if it belongs to P_k. h_k is the max number of blocks that can be part of P_k. Tools like CPLEX can be used to solve the ILP problems. An application of ILP method to solve the HSP problem is given in [8].

14.3.2 Dynamic programming

Dynamic Programming (DP) aims to solve large problems by breaking them down into several smaller problems. The solution of the initial problem is then established by finding the solutions to the individual smaller problems.

　　The DP method consists of three steps: 1) Defining sub-problems, 2) Finding recurrences, 3) Solving the base cases. Examples of using the DP method for solving the HSP problem are described in [9] and [10].

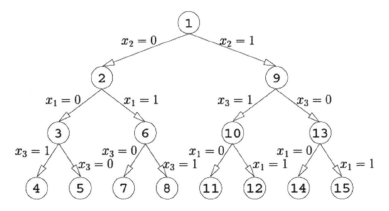

Figure 14.2 Example of binary search tree.

14.3.3 Branch and bound

The method is explained in [11]. The algorithm is based on a binary search tree where each leaf corresponds to a possible solution. At each level of the tree, variables can take the values 1 or 0 as shown in Figure 14.2.

The goal is to find the shortest path from the top to the bottom of the search tree. The shortest path is the path with the minimum cost (objective function) that respects the defined constraints.

The algorithms proposed in [12] and in [13] demonstrate the application of the BB method to deal with the HSP problem.

14.4 Classical Heuristic Approaches

Classical heuristic solutions include the Genetic Algorithm (GA), the Simulated Annealing (SA) algorithm, the Tabu Search (TS) algorithm, the Greedy Algorithm (GR), the Hill Climbing (HC) algorithm, and the Kernighan-Lin (KL) algorithm.

14.4.1 Genetic Algorithm

The GA algorithm consists of the following steps: First, the algorithm generates an initial population (a set of candidate solutions). Second, the algorithm evaluates each individual using a predefined fitness function. Then, a termination test is executed. If the best individual satisfies the termination condition, the algorithm stops and a new population is generated elsewhere using the next steps. Third, the step of selection is executed to select the best parents for each individual. Fourth, the crossover is applied to each parent to generate a new offspring. Fifth, the newly generated offspring is mutated. The elitism principle is applied to keep the individuals with the highest fitness in the newly generated population. The algorithm is repeated from the evaluation step until a terminal condition is met.

The GA algorithm is adopted to solve the HSP problem in several articles ([14, 15, 16, 17, 18, 19, 20, 21, 22]). The proposed approaches used different forms of GA to optimize the overall execution time and the total hardware cost of the system. Numerous approaches deployed a combination of the GA algorithm with another heuristic algorithm. In [23], the GA algorithm is combined with the Tabu Search algorithm. In [24], however, the algorithm combines the genetic GA and the Clustering Algorithm with the objective of minimizing the hardware cost under the constraint of the execution time.

14.4.2 Simulated annealing algorithm

The SA algorithm uses the analogy between the solid annealing principle and the combinatorial optimization problems. First, the algorithm chooses an initial temperature and randomly generates an initial solution. Second, some neighbor solutions are generated. Third, the acceptance function is applied. The latter depends on the energy of both the neighbor and the current solution as well as on the temperature. Fourth, the temperature is reduced depending on a predefined cooling ratio. Then, the algorithm iterates from the second step until a stopping condition are satisfied.

Articles [3] and [25] are examples of the use of the SA algorithm to tackle the HSP problem. In [26], the proposed algorithm combines the SA algorithm with the genetic algorithm.

14.4.3 Tabu search algorithm

Tabu Search initially constructs a first configuration $(x^{now} = Hw_0, Sw_0)$ then it creates a potential list of solutions from the immediate neighbors of x^{now}. Each solution is then evaluated in the third step to end up with the best admissible one. In the last step, if a stopping condition is satisfied, the algorithm terminates, updates the "Tabu" and "Aspiration" conditions, and iterates starting from the second step.

Examples of approaches based on the TS algorithm are presented in [2], [23], and [25]. In [27], the proposed approach is based on a combination of the TS and the SA algorithms. The objective targeted both the partitioning and the scheduling problems on reconfigurable systems (FPGA).

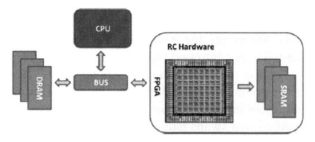

Figure 14.3 Reconfigurable device.

An overview of a reconfigurable system is shown in Figure 14.3. In [28], the proposed solution was based on a combination of the knapsack problem and the TS algorithm. In [29], the proposed solution was based on the combination of the TS and the GA algorithms.

14.4.4 Greedy algorithm

The Greedy Algorithm constructs the solution iteratively. First, it starts with a partial set of elements. Then, at each iteration, the algorithm adds the elements that lead to the best optimization. The actual solution is local with the hope to find a globally optimal solution in the next iterations. The algorithm terminates when no improvement is obtained after a certain number of iterations.

In the algorithm proposed in [30] for dealing with the HSP, in the initial step, all blocks are implemented in hardware. At each step, the algorithm moves a block that minimizes the cost and satisfies the constraints. It also moves the function's successors if these still minimize the cost and satisfy the constraints. In some situations, the algorithm gives only a locally optimal solution, in that case, the global optimal solution is not guaranteed. In [31], the proposed algorithm is an enhancement of the Greedy Algorithm to avoid the locally optimal solution. It utilized a combination of the GR algorithm with two other methods called "constructive method" and "destructive method".

14.4.5 Hill climbing algorithm

Similar to the Greedy Algorithm, *HC* Algorithm starts with a sub-optimal solution and then improves the solution iteratively. The *HC* algorithm also has the advantage of leading to globally optimal solutions and avoiding the local minima. The *HC* algorithm was used in the *HSP*, for example, in [32]. It starts with all blocks in *SW* and repeatedly moves the neighbor block that has the minimum execution time to *HW*.

14.4.6 Kernighan-Lin algorithm

Like the Greedy Algorithm, the Kernighan-Lin method consists of partitioning a graph (G(V; E)) into two separate partitions which are connected optimally. The goal is to regroup the blocks that lead to the biggest gain in cost under constraints on the balance of partition sizes. The algorithm starts with finding an initial partition. Then, at each iteration, it improves by a greedy method that swaps pairs of vertices from the two partitions to minimize the size of the cutset. The algorithm iterates as long as it finds a better partition.

[33], [34] and [35] are examples of proposed approaches based on the KL method to solve the HSP problem.

14.5 Modern Heuristic Approaches

Modern approaches are based on heuristic algorithms such as Particle Swarm Optimization (PSO), Ant Colony Optimization (ACO), Artificial Immune Systems (AIS), the Pareto Optimal optimization, Path-Based optimization, Fuzzy Logic, and other heuristics and meta-heuristics algorithms.

14.5.1 Particle swarm optimization

PSO is described in [36]. The algorithm mimics the work of animal societies with no leader. In fact, it is composed of a swarm of particles (potential solutions). Particles will move through a multidimensional search space to find the optimal position. The steps of the algorithm are as follows:

1.   Step 1. Generates the particles and their velocities (randomly)

2.   Step 2. At each iteration:

   a. Calculates the fitness of all particles

   b. Updates the best fitness for each particle (pbest)

   c. Updates the global best fitness (gbest)

   d. Updates the position and the velocity using specific formulas

3.   Step 3. Stop if the termination condition is satisfied, else re-iterate from Step 2

Articles [37], [38], [39], [40] and [41] are examples of proposed solutions using the *PSO* to deal with the *HSP*.

14.5.2 Ant colony optimization

ACO is described in [42]. The algorithm is inspired by the natural behavior of ants. In fact, ants mark their way down to food using pheromone trails. At first, ants move randomly and depose the pheromone on different parts of the route until they find the right and shortest path that links the food source to their colony. The right path comprises a high level of pheromone, unlike the other paths, which increases the probability of other ants following it. Thus, this behavior leads to the discovery of the shortest paths.

As defined in [43], the working of the ACO algorithm is shown in Figure 14.4.

The steps of the algorithm are as follows:

1. Define the initial parameters

2. Give a correspondence of the pheromone trails to the problem

3. Set the heuristic preference for ants

4. while termination conditions are not met, do:

   a. Explore the search space and construct ant solutions

   b. Probabilistically, choose the next step according to the given phero-mone model

   c. Constructed solutions using local search (Optional)

   d. Update pheromones (add new + evaporate)

5. end while

[44] and [45] are examples of using the *ACO* to solve the *HSP* problem.

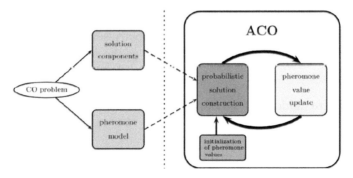

Figure 14.4 The working of the ACO meta-heuristic.

14.5.3 Artificial immune systems

AIS is a branch of artificial intelligence, as defined in [46]. An overview of the immune system (*IS*) is presented in [47]. The role of *IS* is to protect the body from infection. It is comprised of:

1. The immune recognition is based on the complementarily between the binding region of a portion of the antigen (epitope) and the receptor

2. Different antibodies (single type of receptor) can recognize a single antigen

3. The Clonal selection has the following characteristics:

   a. Eliminate self-antigens

   b. Proliferate and differentiate mature lymphocytes with antigen

   c. Restrict one pattern to one differentiated cell and apply the clonal descendants

   d. Generate new genetic changes randomly

4. T-cells have the role of regulation of other cells

5. The primary and the secondary immune responses remember encounters

6. The immune system has to make difference between self and nonself cells

7. As the antigenic encounters may lead to cell death, positive and negative selections are needed

The approach proposed in [48] is a good example of applying AIS to deal with the hardware/software partitioning problem. The objective is to minimize the total hardware cost subject to the execution time constraint. The possible solutions play the role of the antibody set of the immune system, while the partitioning problem is regarded as the antigen. The steps of the algorithm are:

1. Initialize randomly a set of antibodies

2. Initialize the generation: Gen = 0

3. Terminates if the maximal generation is reached

4. Mutation phase: Each antibody will generate an immature antibody

5. Evaluation phase: The affinity of each newly generated immature anti-body is calculated

6. Filter the new antibodies by negative selection

7. Selection phase: N antibodies are selected for the next generation; the selection is applied to the previous generation antibody set and the intermediate set generated in step 6.

8. Update the self-set S.

9. Gen = Gen + 1, and go to step 3.

14.5.4 Pareto optimal

The problem has the formulation described as follows:

$$\min\left(f_1\left(x_1,x_2,\ldots,x_n\right),f_2\left(x_1,x_2,\ldots,x_n\right),\ldots,\ f_m\left(x_1,x_2,\ldots,x_n\right)\right)$$
$$subject\ to: g_k\left(x_1,x_2,\ldots,x_n\right)\leq 0\ ,\ 1\leq k\leq p$$
$$x_i \in \{0,1\},\ 1\leq i\leq n \tag{14.2}$$

The Pareto-dominance is defined by the following: Let X_1 and X_2 solutions to the problem described in 14.2, and X is a set of all possible solutions which satisfy the constraints of 14.2. X_1 is said to dominate X_2 if the conditions described in 14.3 are satisfied.

$$\forall i \in \{1,\ldots,n\},\quad f_i\left(X_1\right)\leq f_i\left(X_2\right)$$
$$\exists j \in \{1,\ldots,n\},\quad f_j\left(X_1\right)<f_j\left(X_2\right) \tag{14.3}$$

The set of solutions that are not dominated by the other solutions is called the set of Pareto-Optimal solutions. In the example of Figure 14.5, the set of Pareto-Optimal is composed of points: {1,2,4,6}.

Based on Pareto Optimal, several algorithms were proposed to deal with the HSP problem. The articles presented in [49], [50], [51], [52], [53] and [54] are examples of such proposed approaches.

4.5.5 Path-based algorithms

Several approaches were based on graph optimization while dealing with the HSP problem. Depending on the model used, the optimization is aimed at either finding the shortest path or minimizing the critical path.

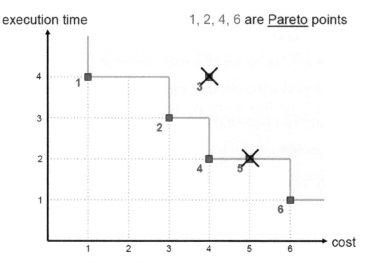

Figure 14.5   Pareto-optimal solutions.

For optimal path optimization, the adopted model for the system's representation is the same as presented in [55]. As shown in Figure 14.6, the basic scheduling blocks (BSB) are directly derived from functional specifications represented as data/control flow graphs. The parent blocks can then be grouped as a single block instead of all the child blocks that composed them.

The corresponding computational model and the HW/SW partitioning problem to find the shortest path in a direct graph, having a unique entry point and unique exit point, are respectively represented in Figure 14.7.a and Figure 14.7.b as used in [56].

By finding the shortest path from the entry to the exit points, the system is automatically split into two sets (HW, SW). A plethora of studies has adopted this model to solve the HSP problem and have used different algorithms for solving the optimal path problem. Examples of those studies are presented in [10], [55], [56], [57], [58], [59], [60] and [61].

For the critical path optimization, the system is modeled as any data acyclic graph (*DAG*). The critical path is the longest path in terms of execution time. The objective is to minimize this critical path and also optimize the other parameters such as the hardware cost and the power consumption. In [62], an algorithm based on a combination of the Shuffled Frog Leaping Algorithm and Greedy Algorithm was proposed to minimize the critical path. In this study, the hardware cost parameter is defined as a constraint. The proposed approaches presented in [63] and [64] have the same objectives as the problem described in [62]. In [63], the execution time of the whole system

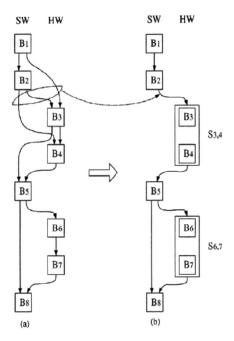

Figure 14.6 Partitioning model. [55]: (a) example of actual data-dependencies between hardware and software blocks, (b) how data dependencies between adjacent hardware blocks and software blocks are interpreted in the model.

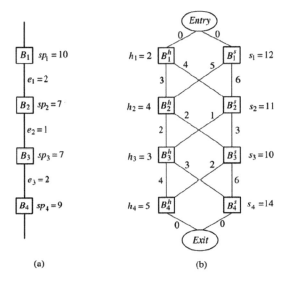

Figure 14.7 Partitioning model: (a) example of a computational model, (b) corresponding graph with unique entry and unique exit points.

is defined as the execution time of the critical path in the graph. In [64], the approach was based on the Brainstorm Optimization Algorithm. Another approach based on critical path optimization is presented in [65].

4.5.6 Fuzzy logic

Fuzzy Logic is described in [66]. The method is grounded upon the fuzzy set theory. This theory is a generalization of the classical Boolean logic. Fuzzy logic has the following characteristics:

1. Consider a set X, and a set A to be is a fuzzy subset of X. A is characterized by a membership function named μ_A

2. $\mu_A : X \rightarrow [0, 1]$. $\mu_A(x)$ is called the membership degree of x in A

3. Let be a variable, is the range of values of the variable, and T_V is a set of fuzzy sets. The triplet (V, X, T_V) corresponds to the linguistic variable

4. The definition of operators (AND, OR, and NOT) on the fuzzy sets is chosen, as the membership functions

5. Fuzzy reasoning is based on fuzzy rules, for example:

    a. *if $x \in A$ and $y \in B$ then $z \in C$*

6. It is noted that there is no single definition of fuzzy implications

The final main step that allows switching from the fuzzy set (which results from the aggregation of the results to one decision) is called defuzzification. A short overview diagram of a fuzzy system is shown in Figure 14.8.

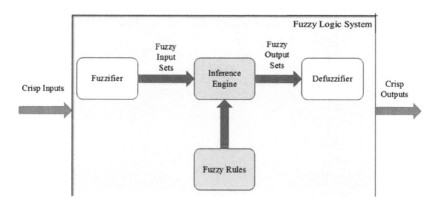

Figure 14.8   Fuzzy logic system structure.

An example of applying the Fuzzy logic to tackle the HSP problem is given in [67]. The goal of this method is to minimize the hardware cost and the execution time simultaneously.

14.5.7 Other heuristic algorithms

Other heuristics and meta-heuristics algorithms were also proposed to solve the HSP problem. In [68], the authors used the Shuffled Frog Leaping Algorithm (SFLA) aiming at minimizing the time cost of the critical path of a DAG graph under a hardware area constraint. Another example of the use of the SFLA algorithm to deal with the HSP problem is presented in [69]. In [70], the authors proposed an approach based on Artificial Bee Colony Optimization (ABC). In [71] and in [72], the proposed algorithm was based on Neural Network optimization. In [73], the proposed approach dealt with the Satisfiability Modulo Theories (SMT) framework. An improved mixed-integer linear programming (MILP) algorithm was proposed in [74]. In [75], the authors used the Multi-Objective Parallel Search (MOPS) algorithm, which is a kind of swarm intelligence manipulating a population of agents. In [76], the proposed solution is a combination of Dynamic Programming and a Greedy Repair algorithm. New solutions, such as the [77] and [78] were also proposed.

14.6 New Perspectives

The objective is to study the HSP problem while taking into consideration different aspects such as the number of metrics involved, the graph used to model the system, and the target software architecture (mono-processor or multiprocessors). The proposed algorithms were grouped into three categories: the first category was based on the Lagrangian relaxation method, the second category was framed according to the GO game, and the third category was based on the Balas method and 0–1 Knapsack algorithm.

14.6.1 First category algorithms

The first category studied the problem in the case of a single-processor system. The latter is modeled as a *DAG* graph in which each block communicates directly with its adjacent block as presented in Figure 14.9.

From the original graph, a developed graph is constructed with the two possible implementations (*HW* or *SW*) of each block as shown in Figure 14.10. Hence, solving the problem is equivalent to finding the optimal in this constructed graph.

Figure 14.9   Example of a simple DAG model.

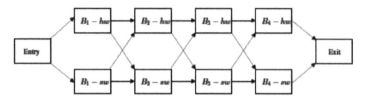

Figure 14.10   Graph representation with four blocks.

First, the goal started by addressing a problem with two parameters: the cost (hardware area) and the performance (execution time). The proposed solution was based on the Lagrangian Relaxation method. Second, the research extended the Lagrangian Relaxation method when three parameters were involved in the *HSP* problem. Finally, the third crucial parameter considered in this study is power consumption.

For the partitioning problem with two parameters, the proposed algorithm in ([79]) aimed at minimizing one parameter concerning a given constraint on the other parameter. This study considered the optimization of the hardware cost under a constraint on the execution time. The experimental results showed that the proposed solution led to more optimal solutions compared to the Genetic Algorithm.

For the partitioning problem with three parameters, the two proposed algorithms aimed at optimizing the system's parameters by minimizing the power consumption concerning the given constraints on the cost parameter and on the execution time parameter. For the first proposed solution ([80]), the Lagrangian Relaxation method was extended by using a combination with the Subgradient method. The empirical results demonstrated that the use of this solution allowed more optimization compared to the Simulated Annealing algorithm and the Genetic Algorithm. However, the disadvantage of this proposed algorithm resides in being quite slow. To eliminate this issue, a second solution was proposed ([81] and [82]). The solution combines the *LR* method with the Genetic algorithm and the Knapsack algorithm. The new experimental results proved that this second solution led approximately to the same results as the first solution on the behalf of system parameters optimization. Moreover, they demonstrated that this solution is very fast, especially when it is compared to the first solution and SA/GA algorithms.

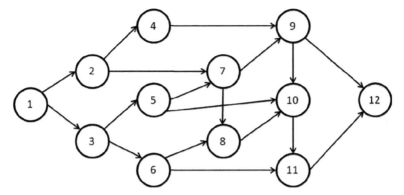

Figure 14.11 Example of DAG representation.

14.6.2 Second category algorithms

The second category considered a system modeled as any data acyclic graph as presented in Figure 14.11.

The first study started with a single-processor system and then the problem in the case of a multiprocessor system was tackled. The proposed solutions were based on artificial intelligence using the MiniMax algorithm and on game theory inspired by the GO game. For a single-processor system, two solutions were proposed: The first algorithm ([83]) allowed the optimization of two parameters: the hardware area (cost) and the execution time (performance). The second algorithm ([84]) allowed the optimization of three parameters by adding the power consumption parameter. For the case of a multi-processor system, an algorithm has been proposed to optimize two parameters: the hardware area (cost) and the performance (execution time). The solution ([85]) is based on the same principle as the first algorithm used in the case of a single-processor system while adding some improvements to take advantage of the existence of several processors. Experimental studies have shown that the proposed algorithms are faster and allowed to obtain better optimizations when compared to the GA and SA algorithms.

14.6.3 Third category algorithms

In the third category, the system was modeled as any data acyclic graph as in Figure 14.11. This part dealt with the problem to minimize more than three metrics of the system. The involved parameters were: the hardware cost, the execution time, the development time (IP reuse and time to market), the power consumption, the quality (maturity of the *HW* and quality of the *SW*), the bandwidth, and the shared memory usage. To achieve this goal, two heuristic approaches with different purposes were proposed. The first approach ([86]) took advantage of the possibility of Balas' method to minimize an objective function under several constraint functions. The study used this method to minimize the hardware area (cost) parameter under the given constraints on the other parameters: the execution time, the power consumption, and the shared memory usage. The second proposed approach ([87]) had the inverse goal. In fact, the approach aimed to optimize several metrics (power consumption, execution time, quality, bandwidth, development time, and shared memory usage) concerning the constraint on one parameter (hardware cost). The algorithm is based on the total profit which is obtained when each block is implemented in *HW*. This profit is calculated using a sum function of each weighted parameter's benefit if a block is implemented in *HW* over its development in *SW*. This solution was developed using the Knapsack algorithm and it was compared to GA and the SA algorithms. Empirical results demonstrated that the KP method is the best choice for the proposed approach as it proved to be faster and to achieve more optimal solutions.

14.7 Conclusion

This chapter studied the hardware/software Partitioning process. First, it presented the problem and then studied the different algorithms used to tackle the aforementioned problem. Some of the proposed algorithms were based on exact algorithms, while the others were based on heuristic and meta-heuristic algorithms. Most of the proposed approaches studied the *HSP* problem with two parameters: the hardware cost and the performance (overall execution time). Some approaches studied the multiobjective optimization by considering the power consumption parameter as well. The new perspectives have studied the partitioning problem while taking into consideration different major aspects such as the criteria (the hardware cost, the execution time, the power consumption, etc.), the graph modeling of the system, and the software architecture (single-processor or multi-processors). The proposed algorithms were based on the Lagrangian relaxation method, game theory, and Balas method. Each of the proposed algorithms was compared to the Simulated Annealing algorithm and the Genetic Algorithm. Different empirical results proved that the proposed algorithms presented fast convergence and led to more optimal solutions. In future work, as the software architecture takes more and more place in system design, the goal would be to study the tasks scheduling problem and the optimization of the access to the shared memory.

References

[1] Shuai Guo Li, Fu Jin Feng, Hua Jun Hu, Cong Wang, and Duo Qi. Hardware/software partitioning algorithm based on genetic algorithm. In Journal of Computers, volume 9. 2014.

[2] Lin Geng, Zhu Wenxing, and Ali Montaz. A tabu search-based memetic algorithm for hardware/software partitioning. In Mathematical Problems in Engineering, pages 1309–1315. 2014.

[3] Sudarshan Banerjee and Nikil Dutt. Very fast simulated annealing for hw-sw partitioning. In Technical Report, CECS-TR-04-17. 2004.

[4] Jorg Henkel. A low power hardware/software partitioning approach for core-based embedded systems. In Proceedings 1999 Design Automation Conference (Cat. No. 99CH36361), pages 122–127. IEEE, 1999.

[5] Wenjun Shi, Jigang Wu, Siew-kei Lam, and Thambipillai Srikanthan. Algorithms for bi-objective multiple-choice hardware/software partitioning. In Computers & Electrical Engineering, volume 50, pages 127–142. 2016.

[6] Edwin Sha, Li Wang, Qingfeng Zhuge, Jun Zhang, and Jing Liu. Power efficiency for hardware/software partitioning with time and area constraints on mpsoc. In International Journal of Parallel Programming, volume 43, pages 381–402. 2015.

[7] Wayne Wolf. A decade of hardware/software codesign. Computer, 36(4):38–43, 2003.

[8] Ralf Niemann and Peter Marwedel. Hardware/software partitioning using integer programming. In Proceedings of the 1996 European Conference on Design and Test, pages 473–, 1996.

[9] P. V. Knudsen and J. Madsen. Pace: a dynamic programming algorithm for hardware/software partitioning. In Proceedings of 4th International Workshop on Hardware/Software Co-Design. Codes/CASHE '96, pages 85–92, 1996.

[10] Jigang Wu and Thambipillai Srikanthan. Low-complex dynamic programming algorithm for hardware/software partitioning. Inf. Process. Lett., 98(2):41–46, 2006.

[11] Jens Clausen. Branch and bound algorithms-principles and examples. Department of Computer Science, University of Copenhagen, pages 1–30, 1999.

[12] Mann Zoltan Adam, Andras Orban, and Peter Arato. Finding optimal hardware/software partitions. In Form. Methods Syst. Des., volume 31, pages 241–263. 2007.

[13] Wu Jigang, Baofang Chang, and Thambipillai Srikanthan. A hybrid branch-and-bound strategy for hardware/software partitioning. In

2009 Eighth IEEE/ACIS International Conference on Computer and Information Science, pages 641–644. IEEE, 2009.

[14] D. Saha, R. S. Mitra, and A. Basu. Hardware software partitioning using genetic algorithm. In Proceedings Tenth International Conference on VLSI Design, pages 155–160, 1997.

[15] Madhura Purnaprajna, Marek Reformat, and Witold Pedrycz. Genetic algorithms for hardware software partitioning and optimal resource allocation. In Journal of Systems Architecture, volume 53, pages 339–354. 2007.

[16] P. Arato, S. Juhasz, Z. A. Mann, A. Orban, and D. Papp. Hardware-software partitioning in embedded system design. In IEEE International Symposium on Intelligent Signal Processing, pages 197–202, 2003.

[17] Chehida, K. Ben, and M. Auguin. Hw/sw partitioning approach for reconfigurable system design. In Proceedings of the 2002 International Conference on Compilers, Architecture, and Synthesis for Embedded Systems, pages 247–251, 2002.

[18] B. Knerr, M. Holzer, and M. Rupp. Novel genome coding of genetic algorithms for the system partitioning problem. In 2007 International Symposium on Industrial Embedded Systems, pages 134–141, 2007.

[19] P. Mudry, G. Zufferey, and G. Tempesti. A hybrid genetic algorithm for constrained hardware-software partitioning. In 2006 IEEE Design and Diagnostics of Electronic Circuits and systems, pages 1–6, 2006.

[20] Pierre-Andr´e Mudry, Guillaume Zufferey, and Gianluca Tempesti. A dynamically constrained genetic algorithm for hardware-software partitioning. In Proceedings of the 8th annual conference on Genetic and evolutionary computation, pages 769–776, 2006.

[21] Shuai Guo Li, Fu Jin Feng, Hua Jun Hu, Cong Wang, and Duo Qi. Hardware/software partitioning algorithm based on genetic algorithm. JCP, 9(6):1309–1315, 2014.

[22] Peter Arato, Sandor Juhasz, Zoltan Adam Mann, Andras Orban, and David Papp. Hardware-software partitioning in embedded system design. In IEEE International Symposium on Intelligent Signal Processing, 2003, pages 197–202. IEEE, 2003.

[23] G Li, J Feng, C Wang, and J Wang. Hardware/software partitioning algorithm based on the combination of genetic algorithm and tabu search. In Engineering Review, volume 34, pages 151–160. 2014.

[24] Li Weijia, Li Lanying, Sun Jianda, Lv Zhiqiang, and Fei Guan. Hardware/software partitioning of combination of clustering algorithm and genetic algorithm. In International Journal of Control and Automation, volume 7, pages 347–356. 2014.

[25] Petru Eles, Zebo Peng, Krzysztof Kuchcinski, and Alexa Doboli. Hardware/software partitioning with iterative improvement heuristics. In Proceedings of the 9th International Symposium on System Synthesis, pages 71–. 1996.

[26] Xibin Zhao, Hehua Zhang, Yu Jiang, Songzheng Song, Xun Jiao, and Ming Gu. An effective heuristic-based approach for partitioning. Journal of Applied Mathematics, 2013, 2013.

[27] Peng Liu, Jigang Wu, and Yongji Wang. Hybrid algorithms for hardware/software partitioning and scheduling on reconfigurable devices. Mathematical and Computer Modelling, 58(1-2):409–420, 2013.

[28] Jigang Wu, Pu Wang, Siew-Kei Lam, and Thambipillai Srikanthan. Efficient heuristic and tabu search for hardware/software partitioning. The Journal of Supercomputing, 66(1):118–134, 2013.

[29] Lanying Li and Min Shi. Software-hardware partitioning strategy using hybrid genetic and tabu search. In 2008 International Conference on Computer Science and Software Engineering, volume 4, pages 83–86. IEEE, 2008.

[30] MC Bhuvaneswari and M Jagadeeswari. Hardware/software partitioning for embedded systems. In Application of Evolutionary Algorithms for Multi-objective Optimization in VLSI and Embedded Systems, pages 21–36. 2015.

[31] G. Lin. An iterative greedy algorithm for hardware/software partitioning. In 2013 Ninth International Conference on Natural Computation (ICNC), pages 777–781. 2013.

[32] Joon Edward Sim, Tulika Mitra, and Weng-Fai Wong. Defining neighborhood relations for fast spatial-temporal partitioning of applications on reconfigurable architectures. In Proceedings of International Conference on ICECE Technology, pages 121 – 128, 2009.

[33] Frank Vahid. Modifying min-cut for hardware and software functional partitioning. In Proceedings of 5th International Workshop on Hardware/Software Co Design. Codes/CASHE'97, pages 43–48. IEEE, 1997.

[34] Frank Vahid and Thuy Dm Le. Extending the kernighan/lin heuristic for hardware and software functional partitioning. Design automation for embedded systems, 2(2):237–261, 1997.

[35] ZoltaN Mann, Andras Orban, and Viktor Farkas. Evaluating the kernighan-lin heuristic for hardware/software partitioning. International Journal of Applied Mathematics and Computer Science, 17(2):249–267, 2007.

[36] dian Palupi Rini, Siti Mariyam Shamsuddin, and Siti Yuhaniz. Particle swarm optimization: Technique, system and challenges. International Journal of Computer Applications, 1, 09 2011.

[37] Amin Farmahini Farahani, Mehdi Kamal, Seid Mehdi Fakhraie, and Saeed Safari. hw/sw partitioning using discrete particle swarm. In Proceedings of the 17th ACM Great Lakes symposium on VLSI, pages 359–364,2007.

[38] MB Abdelhalim and SED Habib. Particle swarm optimization for hw/sw partitioning. Particle Swarm Optimization, pages 49–76, 2009.

[39] Mohamed B Abdelhalim, AE Salama, and SE-D Habib. Constrained and unconstrained hardware-software partitioning using particle swarm optimization technique. In Embedded System Design: Topics, Techniques and Trends, pages 207–220. Springer, 2007.

[40] Xiaohu Yan, Fazhi He, Neng Hou, and Haojun Ai. An efficient particle swarm optimization for large-scale hardware/software co-design system. International Journal of Cooperative Information Systems, 27(01):1741001, 2018.

[41] Alakananda Bhattacharya, Amit Konar, Swagatam Das, Crina Grosan, and Ajith Abraham. Hardware software partitioning problem in embedded system design using particle swarm optimization algorithm. In 2008 International Conference on Complex, Intelligent and Software Intensive Systems, pages 171–176. IEEE, 2008.

[42] Marco Dorigo, Mauro Birattari, and Thomas Stutzle. Ant colony optimization. IEEE computational intelligence magazine, 1(4):28–39, 2006.

[43] Christian Blum. Ant colony optimization: Introduction and recent trends. Physics of Life Reviews, 2(4):353–373, 2005.

[44] Fabrizio Ferrandi, Pier Luca Lanzi, Christian Pilato, Donatella Sciuto, and Antonino Tumeo. Ant colony optimization for mapping, scheduling and placing in reconfigurable systems. In 2013 NASA/ESA Conference on Adaptive Hardware and Systems (AHS-2013), pages 47–54. IEEE, 2013.

[45] Gang Wang, Wenrui Gong, and Ryan Kastner. Application partitioning on programmable platforms using the ant colony optimization. journal of Embedded Computing, 2(1):119–136, 2006.

[46] Leandro Nunes Castro, Leandro Nunes De Castro, and Jonathan Timmis. Artificial immune systems: a new computational intelligence approach. Springer Science & Business Media, 2002.

[47] S. Kellie and Z. Al-Mansour. Chapter four - overview of the immune system. In Mariusz Skwarczynski and Istvan Toth, editors, Micro and Nanotechnology in Vaccine Development, pages 63–81. William Andrew Publishing, 2017.

[48] Yiguo Zhang, Wenjian Luo, Zeming Zhang, Bin Li, and Xufa Wang. A hardware/software partitioning algorithm based on artificial immune principles. Applied Soft Computing, 8(1):383–391, 2008.

[49] Jiang Hong, Yang Meng-fei, Zhang Shao-lin, and Wang Ruo-chuan. A new method for multi-objective optimization problem. In 2013 IEEE 4th International Conference on Electronics Information and Emergency Communication, pages 209–212. IEEE, 2013.

[50] Cagkan Erbas, Selin C Erbas, and Andy D Pimentel. A multiobjective optimization model for exploring multiprocessor mappings of process networks. In Proceedings of the 1st IEEE/ACM/IFIP international conference on Hardware/software codesign and system synthesis, pages 182–187, 2003.

[51] M Jagadeeswari and MC Bhuvaneswari. Efficient multiobjective genetic algorithm for hardware-software partitioning in embedded system design: Enga. International journal of computer applications in technology, 36(3-4):181–190, 2009.

[52] Yang Liu and Qing Cheng Li. Hardware software partitioning using immune algorithm based on pareto. In 2009 International Conference on Artificial Intelligence and Computational Intelligence, volume 2, pages 176–180. IEEE, 2009.

[53] Cagkan Erbas, Selin Cerav-Erbas, and Andy D Pimentel. Multiobjective optimization and evolutionary algorithms for the application mapping problem in multiprocessor system-on-chip design. IEEE Transactions on Evolutionary Computation, 10(3):358–374, 2006.

[54] Anup Kumar Das, Akash Kumar, Bharadwaj Veeravalli, and Francky Catthoor. Reliability and energy-aware codesign of multiprocessor systems. In Reliable and Energy Efficient Streaming Multiprocessor Systems, pages 75–101. Springer, 2018.

[55] Jan Madsen, Jesper Grode, Peter Voigt Knudsen, Morten Elo Petersen, and Anne Haxthausen. Lycos: The lyngby co-synthesis system. Design Automation for Embedded Systems, 2(2):195–235, 1997.

[56] Ji-Gang Wu, Thambipillai Srikanthan, and Guang-Wei Zou. New model and algorithm for hardware/software partitioning. Journal of Computer Science and Technology, 23(4):644–651, 2008.

[57] Wu Jigang and Srikanthan Thambipillai. A branch-and bound algorithm for hardware/software partitioning. In Proceedings of the Fourth IEEE International Symposium on Signal Processing and Information Technology, 2004., pages 526–529. IEEE, 2004.

[58] Jigang Wu, Thambipillai Srikanthan, and Chengbin Yan. Algorithmic aspects for power-efficient hardware/software partitioning. Mathematics and Computers in Simulation, 79(4):1204–1215, 2008.

[59] Wenjun Shi, Jigang Wu, Siew-kei Lam, and Thambipillai Srikanthan. Algorithms for bi-objective multiple-choice hardware/software partitioning. Computers & Electrical Engineering, 50:127–142, 2016.

[60] Wu Jigang and Thambipillai Srikanthan. Algorithmic aspects of area-efficient hardware/software partitioning. The Journal of Supercomputing, 38(3):223–235, 2006.

[61] Jigang Wu, Qiqiang Sun, and Thambipillai Srikanthan. Algorithmic aspects for multiple-choice hardware/software partitioning. Computers & Operations Research, 39(12):3281–3292, 2012.

[62] T. Zhang, X. Zhao, and X. Li. Efficient hardware/software partitioning based on a hybrid algorithm. IEEE Access, 6:60736–60744, 2018.

[63] Aijia Ouyang, Xuyu Peng, Jing Liu, and Ahmed Sallam. Hardware/software partitioning for heterogenous mpsoc considering communication overhead. International Journal of Parallel Programming, 45, 10 2016.

[64] Tao Zhang, Changfu Yang, and Xin Zhao. Using improved brainstorm optimization algorithm for hardware/software partitioning. Applied Sciences, 9:866, 02 2019.

[65] Nikolina Frid and Vlado Sruk. Critical path method based heuristics for mapping application software onto heterogeneous mpsocs. In 2014 37th International Convention on Information and Communication Technology, Electronics and Microelectronics (MIPRO), pages 1030–1034. IEEE, 2014.

[66] SN Sivanandam, Sai Sumathi, SN Deepa, et al. Introduction to fuzzy logic using MATLAB, volume 1. Springer, 2007.

[67] Humberto Dıaz Pando, Sergio Cuenca Asensi, Roberto Sepulveda Lima, Jenny Fajardo Calderın, and Alejandro Rosete Su´arez. An application of fuzzy logic for hardware/software partitioning in embedded systems. Computacion y Sistemas, 17(1):25–39, 2013.

[68] Jiayi Dua, Xiangsheng Kongb, Xin Zuo, Lingyan Zhangd, and Aijia Ouyange. Shuffled frog leaping algorithm for hardware/software partitioning. Journal of Computers, 9(11):2752–2761, 2014.

[69] Tao Zhang, Xin Zhao, Xinqi An, Haojun Quan, and Zhichun Lei. Using blind optimization algorithm for hardware/software partitioning. IEEE Access, 5:1353–1362, 2017.

[70] Mouloud Koudil, Karima Benatchba, Amina Tarabet, and El Batoul Sahraoui. Using artificial bees to solve partitioning and scheduling problems in codesign. Applied Mathematics and Computation, 186(2):1710 – 1722, 2007.

[71] Tianyi Ma, Xinglan Wang, and Zhiqiang Li. Neural network optimization for hardware-software partitioning. In First International Conference on Innovative Computing, Information and Control-Volume I (ICICIC'06), volume 3, pages 423–426. IEEE, 2006.

[72] Hokchhay Tann, Soheil Hashemi, R Iris Bahar, and Sherief Reda. Hardware-software codesign of accurate, multiplier-free deep neural networks. In 2017 54th ACM/EDAC/IEEE Design Automation Conference (DAC), pages 1–6. IEEE, 2017.

[73] Mingxuan Yuan, Xiuqiang He, and Zonghua Gu. Hardware/software partitioning and static task scheduling on runtime reconfigurable fpgas using a smt solver. In 2008 IEEE Real-Time and Embedded Technology and Applications Symposium, pages 295–304. IEEE, 2008.

[74] Yuchun Ma, Jinglan Liu, Chao Zhang, and Wayne Luk. Hw/sw partitioning for region-based dynamic partial reconfigurable fpgas. In 2014 IEEE 32nd International Conference on Computer Design (ICCD), pages 470–476. IEEE, 2014.

[75] Ihsen Alouani, Braham L Mediouni, and Smail Niar. A multi-objective approach for software/hardware partitioning in a multi-target tracking system. In 2015 International Symposium on Rapid System Prototyping (RSP), pages 119–125. IEEE, 2015.

[76] Naman Govil and Shubhajit Roy Chowdhury. Gma: a high speed metaheuristic algorithmic approach to hardware software partitioning for low-cost socs. In 2015 International Symposium on Rapid System Prototyping (RSP), pages 105–111. IEEE, 2015.

[77] Mourad, Khetatba & Boudour, Rachid. (2021). A Modified Binary Firefly Algorithm to Solve Hardware/Software Partitioning Problem. Informatica. 45. 10.31449/inf.v45i7.3408.

[78] Xian, Tiong & Halim, Zaini & Leong, Ching & Gim, Tan. (2021). Hardware-software partitioning using three-level hybrid algorithm for system-on-chip platform. Bulletin of Electrical Engineering and Informatics. 10. 466-473. 10.11591/eei.v10i1.2201.

[79] Adil Iguider, Mouhcine Chami, Oussama Elissati, and Abdeslam En-Nouaary. Embedded systems hw/sw partitioning based on lagrangian relaxation method. In Proceedings of the Mediterranean Symposium on Smart City Applications, pages 149–160. Springer, 2017.

[80] Adil Iguider, Kaouthar Bousselam, Abdeslam En-Nouaary, Oussama Elissati, and Mouhcine Chami. A novel approach for hardware software partitioning in embedded systems. In 2019 International Conference on Wireless Technologies, Embedded and Intelligent Systems (WITS), pages 1–5. IEEE, 2019.

[81] Adil Iguider, Oussama Elissati, Mouhcine Chami, and Abdeslam En-Nouaary. An efficient hw/sw partitioning algorithm for power optimization in embedded systems. In 2018 International Symposium on

Advanced Electrical and Communication Technologies (ISAECT), pages 1–5. IEEE, 2018.

[82] Adil Iguider, Kaouthar Bousselam, Oussama Elissati, Mouhcine Chami, and Abdeslam En-Nouaary. Heuristic algorithms for multi-criteria hardware/software partitioning in embedded systems codesign. Computers & Electrical Engineering, 84:106610, 2020.

[83] Adil Iguider, Kaouthar Bousselam, Oussama Elissati, Mouhcine Chami, and Abdeslam En-Nouaary. Embedded systems hardware software partitioning using minimax algorithm. In Proceedings of the 4th International Conference on Smart City Applications, pages 1–6, 2019.

[84] Adil Iguider, Kaouthar Bousselam, Oussama Elissati, Mouhcine Chami, and Abdeslam En-Nouaary. Embedded systems hardware software partitioning approach based on game theory. In The Proceedings of the Third International Conference on Smart City Applications, pages 542–555. Springer, 2019.

[85] Adil Iguider, Kaouthar Bousselam, Oussama Elissati, Mouhcine Chami, and Abdeslam En-Nouaary. Go game inspired algorithm for hardware software partitioning in multiprocessor embedded systems. Computer and Information Science, 12(4):111–122, 2019.

[86] Adil Iguider, Abdeslam En-Nouaary, et al. Hw/sw partitioning algorithms for multi-objective optimization in embedded systems. International Journal of Information Science and Technology, 2(2):19–28, 2019.

[87] Adil Iguider, Oussama Elissati, Abdeslam En-Nouaary, and Mouhcine Chami. Heuristic approach for multiobjective hardware/software partitioning. In 2018 19th IEEE Mediterranean Electrotechnical Conference (MELECON), pages 209–212. IEEE, 2018.

Index

About the Editor

Saad Motahhir (https://orcid.org/0000-0002-6846-8908)(Eng., Ph.D., IEEE Senior Member) has previous expertise acting in the industry as Embedded System Engineer at Zodiac Aerospace morocco from 2014 to 2019, and more recently became a professor at ENSA, SMBA University, Fez, Morocco since 2019. He received an engineer degree in the embedded systems from ENSA Fez in 2014. He received his Ph.D. Degree in Electrical Engineering from SMBA University in 2018. He has published a good number of papers in journals and conferences in the last few years, most of which are related to photovoltaic (PV) solar energy and Embedded Systems. He published a number of patents in the Morocco patent office. He edited different books and acted as a guest editor of different special issues and topical collections. He is a reviewer and on the editorial board of different journals. He was associated with more than 30 international conferences as a Program Committee/Advisory Board/Review Board member.

9788770227728